LIQUID CRYSTALS
Basic Handbook

LIQUID CRYSTALS
Basic Handbook

by
Tejas Chadha

Wordpen Academics
Publishing Science That Matter

ISBN: 978-93-6084-583-4

Published by : **WORDPEN ACADEMICS**
4736/23, Ansari Road, Darya Ganj,
New Delhi – 110 002
Phone: +91-011-35004336
E-mail: info@wordpenacademics.com

Printed at **:** **Replika Press Pvt. Ltd.**

Exclusive Distributor: ***Bio-Green Books, New Delhi***

Preface

Liquid crystals represent a fascinating state of matter that exhibits properties of both liquids and solids, giving rise to a rich variety of physical phenomena and practical applications. From the liquid crystal displays (LCDs) that dominate modern electronic devices to the intricate structures found in biological systems, liquid crystals play a crucial role in diverse fields ranging from materials science and physics to biology and technology. As we delve into the world of liquid crystals, we uncover a wealth of knowledge and opportunities for exploration at the interface of chemistry, physics, and engineering.

This book, titled "Liquid Crystals: Basic Handbook," serves as an introductory guide to understanding the fundamental principles, properties, and applications of liquid crystals. Drawing upon the expertise of researchers, educators, and industry professionals in the field, this volume offers insights, explanations, and practical knowledge for students, enthusiasts, and practitioners interested in exploring the intriguing world of liquid crystals.

From the molecular structure and phase behavior of liquid crystals to the optical, electrical, and mechanical properties that make them unique, "Liquid Crystals: Basic Handbook" covers a wide range of topics essential for understanding the science and technology of liquid crystals. Through clear explanations, illustrative diagrams, and real-world examples, readers will gain a solid foundation in the principles governing liquid crystal behavior and the techniques used to characterize and manipulate these materials.

Moreover, this book explores the diverse applications of liquid crystals in various industries and technologies, from displays and optical devices to sensors, coatings, and biomedical applications. By showcasing the versatility and potential of liquid crystals in real-world contexts, we aim to inspire readers to explore new avenues of research and innovation in this dynamic field.

As we embark on this journey of exploration and discovery, it is our hope that "Liquid Crystals: Basic Handbook" will serve as a valuable resource for students, researchers, and professionals seeking to expand their knowledge and understanding of liquid crystals. Whether you are a novice exploring the fundamentals of liquid crystal science or an experienced practitioner seeking to stay abreast of the latest developments in the field, this book invites you to delve into the fascinating world of liquid crystals and unlock new insights into their properties, behaviors, and applications.

Tejas Chadha

Contents

1

Introductory Chapter: Liquid Crystals and Applications

1. Introduction

Liquid crystalline state is the one having the physical properties that fall in between conventional crystalline (solid) and isotropic fluid (liquid). Liquid crystals have been of great interest owing to the amazing physical and chemical properties these exhibit, thereby proving to be potentially useful in versatile technological applications [1]. Though liquid crystals may flow like a liquid, the molecular orientations of them may be of the kind that a solid crystal possesses. Furthermore, liquid crystals may exist naturally or these can also be synthesized [2, 3]. Within the context, the lyotropic phase of liquid crystal can be abundantly found in living organisms, such as proteins and cell membranes. Synthesized forms of liquid crystals are widely used in display applications [4]. The use of liquid crystals in display technology stems from the nature of chirality and the high electro-optic coefficient of these mediums, thereby making them significantly advantageous. Apart from display-related applications, liquid crystals are greatly attractive for several kinds of sensing applications as well. Some other applications would be in the areas of lasers [5] and medical diagnostics [6]. However, the flat panel display is the most recognized device where liquid crystals are widely used. Apart from the display panels, there are a host of other avenues where the synthesized versions of liquid crystals are indispensable.

2. Liquid crystal types and properties

As reported vastly, the nematic, smectic, and cholesteric are the three basic phases of liquid crystals classified according to the molecular orientations [7]. Apart from this, liquid crystals are also technically classified into the thermotropic [3, 8] and lyotropic categories [9], distinguished by the mechanisms responsible for their self-organization. These two kinds of liquid crystals generally do have some similarities in their physical and chemical properties, in spite of the fact that the chemical structures of their molecules greatly differ. The thermotropic class of liquid crystals undergo thermally induced transitions to the liquid crystalline state [3]. As such, raising the temperature of a solid and/or lowering the temperature of a liquid can result in a thermotropic kind of liquid crystals. On the other hand, the lyotropic class of liquid crystals exhibits solvent-induced transitions [9], and therefore, they are amphiphilic in nature. That is, such liquid crystals are composed of both the lyophilic (solvent-attracting) and lyophobic (solvent-repelling) kinds of mediums.

Describing the nematic, smectic, and cholesteric phases of liquid crystals, as stated before, they are categorized by the kind of molecular orientation they possess. In the nematic state, the molecules acquire no positional order, but they are aligned along the director, which represents the direction of molecular orientation. Thus, the molecules in the nematic phase tend to be parallel, but their positions are just random. In the smectic phase, the liquid crystal molecules show a translational order, which is not found in the nematic phase. Therefore, the molecules in this state are aligned parallel, but their centers are stacked in parallel layers within which they acquire random positions. This essentially indicates the restriction imposed over the motion of liquid crystal molecules in the smectic phase, which is confined to the planes in which they are aligned. The molecules of liquid crystals in the cholesteric phase show helicoidal orientation, which makes this phase as a distorted kind of the nematic state, wherein the molecular orientation undergoes helical rotation about an axis. Clearly, the director in the cholesteric phase of liquid crystals exhibits a continuous helical pattern about the axis perpendicular to the two-dimensional nematic layers (of molecules). Within the context, chirality of cholesteric liquid crystals is of great potential in developing many electro-optic components, such as sensors and display panels.

Liquid crystals are anisotropic in nature, that is, these exhibit directional properties. This makes them acquire the birefringence characteristic [10]. Therefore, in liquid crystals, the light-polarized parallel to the director experiences a different refractive index than that polarized perpendicular to the director. This results in the propagation of light waves (in these mediums) in two different directions with different phase velocities, and the respective propagating waves are called the fast and the slow waves (or the ordinary and the extraordinary waves, respectively). Because of splitting of the incoming wave, the ordinary and extraordinary waves propagate with different phase velocities in the birefringent medium, and therefore, the waves become out of phase. Furthermore, the anisotropic characteristic of liquid crystals depends on the ambient temperature, and therefore, the property of birefringence (in liquid crystals) can be eliminated at the nematic-to-isotropic phase transition owing to the altering thermal ambience.

3. Applications of liquid crystals

Due to their possessing very high electro-optic coefficients, they are of great potential in many photonic applications [11]. However, till today liquid crystals find the major applications in flat panel electronic displays (or the LCDs) [12] as these offer several advantages over traditional displays comprising the cathode ray tubes (CRTs). Some of the notable advantages of LCD would be low power consumption and significantly less weight. At the same time, LCDs suffer from the drawback of limited viewing angle and relatively shorter lifetime.

Apart from the LCD panels, liquid crystals are used in many other photonic applications. For example, the chiral nematic phase of liquid crystal can selectively reflect light, which essentially depends on the wavelength of light and the pitch of the helix that the director assumes. To be more specific, if the wavelength of the incoming light is equal to the helix pitch, it will be reflected by the medium. This way, the cholesteric phase of liquid crystals can be used to develop optical filters and imaging systems [13, 14]. Since the cholesteric state is highly temperature dependent, variation in thermal ambience would modify the orientation of director between the successive layers (of the liquid crystal). This, in turn, modifies the length of helix pitch of the director, thereby altering the selectivity of the reflection spectrum. This operation can be utilized in devising thermal sensors [15].

Because of the large electro-optic coefficient of liquid crystals, these are of great potential in electromagnetic field sensing [7]. Upon applying an electric field, the director of the liquid crystal structure would be aligned along the applied field. In photonic applications, these are of special mention in the context of evanescent field sensing as the use of liquid crystals makes the field confinement in the outermost section of a liquid crystal clad-based fiber/guide very large, which can be affected in the presence of a measurand [16]. Within the context, the radially anisotropic liquid crystals have been greatly dealt with in the literature focusing on the analyses of the light wave propagation through optical waveguides composed of such mediums and having different kinds of geometrical features [17, 18, 19]. Liquid crystal-based guides with conducting sheath and tape helixes are new of their kinds, which provide control over the dispersion properties, thereby allowing the on-demand kinds of applications [20, 21]. It must be noted that the liquid crystal-based guides have promising potential in developing sensors in the electrical and chemical industries.

4. Summary

Liquid crystals are of great potential in varieties of technology-oriented applications. Apart from the most recognized usage in the development of display panels, these are also useful in many other applications that include optical imaging and recording, erasable optical disks, electronic slides, light modulators, lasers, etc. Research reports also indicate the use of liquid crystals in biomedical applications and metamaterial-based reconfigurable antennas. R&D scientists have been involved in investigating the properties of varieties of artificially synthesized liquid crystals with the positive thought of finding prudent and novel applications of these.

Finally, the book provides the basic understanding of liquid crystals and also the advanced discussions on phase transitions, which would make a fairly good platform for applications of liquid crystals in developing sensors. It is believed the volume would be useful to the undergraduate students in framing their research topics in the relevant direction.

References

1. Chandrasekhar S. Liquid Crystals. Cambridge: Cambridge University Press; 1992
2. Collings PJ. Liquid Crystals: Nature's Delicate Phase of Matter. Princeton: Princeton University Press; 1990
3. de Gennes PG, Prost J. The Physics of Liquid Crystals. Oxford: Clarendon Press; 1993
4. O'Mara WC. Liquid Crystal Flat Panel Displays. New York: Van Nostrand Reinhold; 1993
5. Coles H, Morris S. Liquid-crystal lasers. Nature Photonics. 2010;4:676-685
6. Lin Y-H. Liquid crystals for bio-medical applications. In: Lee C-C, editor. The Current Trends of Optics and Photonics, Topics in Applied Physics. Vol. 129. Dordrecht: Springer; 2015
7. Choudhury PK, editor. New Developments in Liquid Crystals and Applications. New York: Nova; 2013
8. Shao Y, Zerda TW. Phase transitions of liquid crystal PAA in confined geometries. Journal of Physical Chemistry B. 1998;102:3387-3394
9. Liang Q, Liu P, Liu C, Jian X, Hong D, Li Y. Synthesis and properties of lyotropic liquid crystalline copolyamides containing phthalazinone moiety and ether linkages. Polymer. 2005;46:6258-6265
10. Madsen LA, Dingemans TJ, Nakata M, Samulski ET. Thermotropic biaxial nematic liquid crystals. Physical Review Letters. 2004;92:145505-1-145505-4
11. Moreno I. Liquid crystals for photonics. Optical Engineering. 2011;50:081201-1
12. Castellano JA. Liquid Gold: The Story of Liquid Crystal Displays and the Creation of an Industry. Singapore: World Scientific; 2005
13. Gebhart SC, Stokes DL, Vo-Dinh T, Mahadevan-Jansen A. Instrumentation considerations in spectral imaging for tissue demarcation: Comparing three methods of spectral resolution. Proceedings of SPIE. 2005;5694:41-52
14. Levenson RM, Lynch DT, Kobayashi H, Backer JM, Backer MV. Multiplexing with multispectral imaging: From mice to microscopy. ILAR Journal. 2008;49:78-88
15. Plimpton RG. Pool Thermometer. U.S. Patent 4738549. 1988

16. Choudhury PK. Liquid crystal optical fibers for sensing applications. Proceedings of SPIE. 2013;8818:88180E-1-88180E-10
17. Choudhury PK, Yoshino T. TE and TM modes power transmission through liquid crystal optical fibers. Optik. 2011;115:49-56
18. Choudhury PK, Soon WK. On the transmission by liquid crystal tapered optical fibers. Optik. 2011;122:1061-1068
19. Ghasemi M, Choudhury PK. Propagation through complex structured liquid crystal optical fibers. Journal of Nanophotonics. 2014;8:083997-1-083997-13
20. Ghasemi M, Choudhury PK. On the conducting sheath double-helix loaded liquid crystal optical fibers. Journal of Electromagnetic Waves and Applications. 2015;29:1580-1592
21. Ghasemi M, Choudhury PK. Conducting tape helix loaded radially anisotropic liquid crystal clad optical fiber. Journal of Nanophotonics. 2015;9:093592-1-0093592-15

2

Overview of Liquid Crystal Research: Computational Advancements, Challenges, Future Prospects and Applications

1. Introduction

Liquid crystals investigation was begun by an Austrian scientist, Friedrich Reinitzer in 1888 when he espied a material named cholesteryl benzoate having two different melting points. He manipulated the temperature of the samples in his experiments and observed the changes that happened. Due to this early work, Reinitzer is attributed with the discovery of a new phase of matter - the liquid crystal phase and since then, liquid crystals have been under study for various research fields including physics, chemistry, medicine as well as engineering that resulted in the advanced nanostructured liquid crystals exhibiting an extraordinary ordered pattern and highlighting advanced functions such as electro-optical effects, chromism, sensing and templating [1, 2, 3].

The state of matter which shows liquid and crystalline properties at the same time is known as a liquid crystal, and this dual state of matter not only possesses viscous properties like fluidity, formation of droplets and mechanical properties but also exhibits the nature of crystalline solids such as periodic and anisotropic properties; therefore, they are named mesophases [4]. Liquid crystals are gaining popularity due to their remarkable functional properties, due to their ability to respond to external stimuli such as heat, light, electric fields, and mechanical forces [5]. Liquid phase molecules have no defined intrinsic order, while solid molecules are highly ordered, and the distinguishing property of liquid crystal is the inclination of molecules to point along the same direction of axis, called an anisotropic property [6] as shown in Figure 1 [7]. Liquid crystals are classified into two primary groups based on various conditions and parameters: thermotropic and lyotropic liquid crystals, which differ in their arrangement mechanism but share many properties.

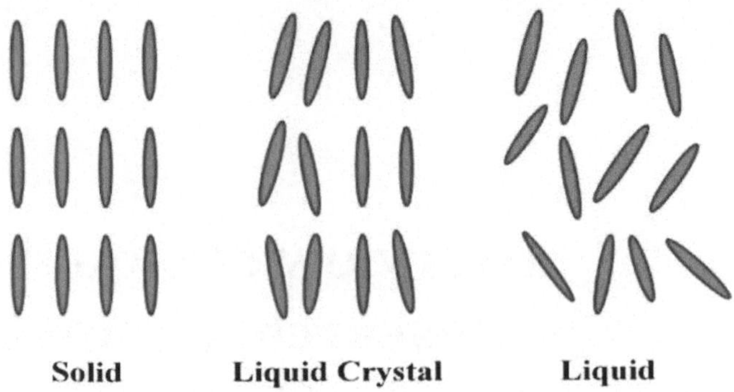

Figure 1. A comparison of solid, liquid and liquid crystal orientation [7].

Thermotropic liquid crystals may be created by heating crystalline structures, which causes molecular position order to disappear but not the molecular orientational order, resulting in a dynamic phase with anisotropic liquid crystal characteristics [8]. This type of liquid crystal is temperature-dependent, with high temperatures resulting in an ordinary isotropic liquid phase owing to disrupted phase ordering, and low temperatures resulting in an anisotropic ordered crystalline liquid phase. On the other hand, an organized structure in a colloidal solution creates lyotropic liquid crystals that are formed by amphiphilic molecules with hydrophilic and hydrophobic portions. By changing the concentration and temperature of amphiphilic molecules, different lyotropic phases may be produced [9].

2. Fundamentals of liquid crystals

Liquid crystal classifications on an extensive level open up ways for various fundamental meso-phases. The chemical and physical properties of liquid crystals vary by changing temperature, solution concentration and other parameters that resulted in long-range orientational as well as positional order of molecules. Thermotropic liquid crystals result in different kinds of meso-phases due to varying temperatures, and these transactions are controlled thermally. The mechanism further divides thermotropic liquid crystals into two categories: enantiotropic liquid crystals, which can be obtained by only decreasing the liquid temperature or increasing the solid temperature, and monotropic liquid crystals, which can be obtained by either increasing the temperature of a solid-state material or decreasing the temperature of a liquid state material to form an aligned as well as ordered liquid crystal arrangement. Depending on the molecular arrangement and symmetry, thermotropic liquid crystals are subdivided into four main categories; nematics, cholesterics, smectics, and columnar liquid crystals. The orientational order, positional order, and bond orientational order are some of the characteristics that describe liquid crystal formations. Nematic mesophase is the fundamental liquid crystal phase in which the molecular orientation is correlated with local molecular-axis orientation that can be controlled by external factors like; an electric field that helps the molecular axis to align accordingly [10], as is shown in Figure 2(a) [11]. These liquid crystal structures have a long-range orientational order but they lack in order of molecular center of mass positions, and are anisotropic having a rigid molecular backbone that outlines the long axis of molecules. Because they have a strong dipole moment and a high refractive index, these crystalline compounds may easily be polarized, making them useful in display applications [12].

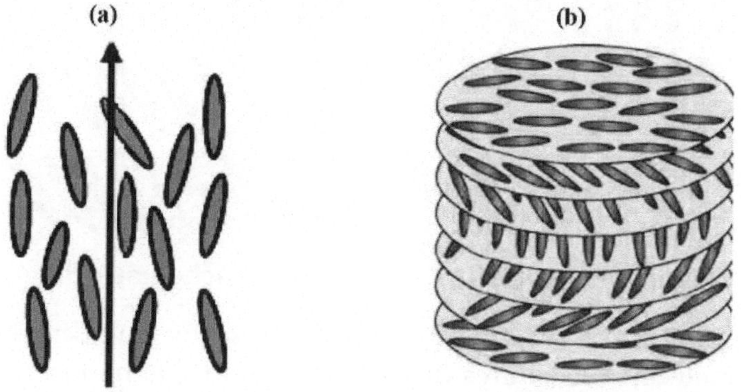

Figure 2. (a) Nematic liquid crystal [11], (b) cholesteric liquid crystal schematic diagram [14].

Cholesteric liquid crystal state is comparable to nematic state [13], having nematic layers where each layer has its director; therefore, it is known as the counterpart of the nematic phase, as is shown in Figure 2(b) [14]. Likewise, these compounds exhibit long-range orientational order owing to their molecular organization into a long axis that is parallel to the molecular plan, and the molecular axis is defined as helical. The cholesteric structure is characterized by the pitch that is the distance along the long-axis on which the director rotates. Cholesterics are finite pitch structures with a few hundred nanometers of pitch; however, nematics can also be of the finite pitch if doped with an enantiomorphic compound. Cholesteric structure pitch is comparable to visible light, resulting in white light scattering owing to Bragg's reflection phenomenon, which may be exploited for different optical applications although this pitch can also be sensitive to numerous variables [15, 16, 17].

Smectic liquid crystals differ from others owing to layer ordering but molecule orientation is comparable to nematic in addition to layer alignment, and these structures have both positional and orientational orders. The smectic liquid crystal phases exhibit molecules organized perpendicular to layers, hexagonally structured, and tilt angle measurement in smectic types A, B, and C; however, they are more ordered than the nematic liquid crystal phases. Figure 3 depicts smectic A and C with no molecular positional ordered inside each individual layer, where smectic A molecules are orientated perpendicular along with the layer, but smectic

C molecules are tilted away from the layer orientation, and so have order in one direction only.

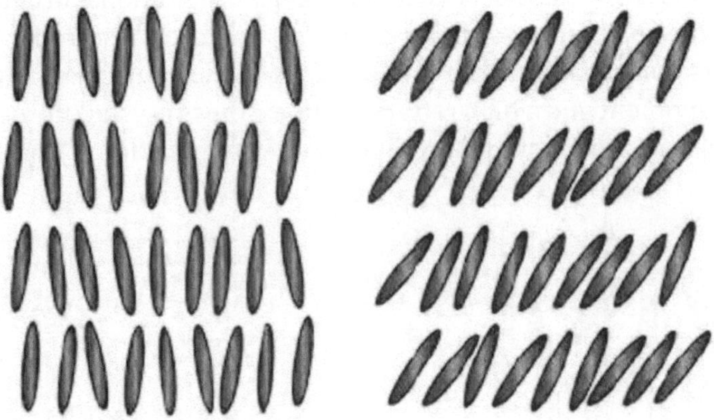

Figure 3. Smectic A, Smectic C liquid crystal schematic diagram [12].

Columnar liquid crystals are also known as discotic liquid crystals due to the disc-shaped molecular structures that are stacked in column form and organized in various shapes [18], while the columnar phase molecules are classified based on their packing motivations. They are arranged in 2D patterns such as hexagonal, rectangular, oblique, and lamellar [19] and these molecules are stacked in their respective lattices as shown in Figure 4.

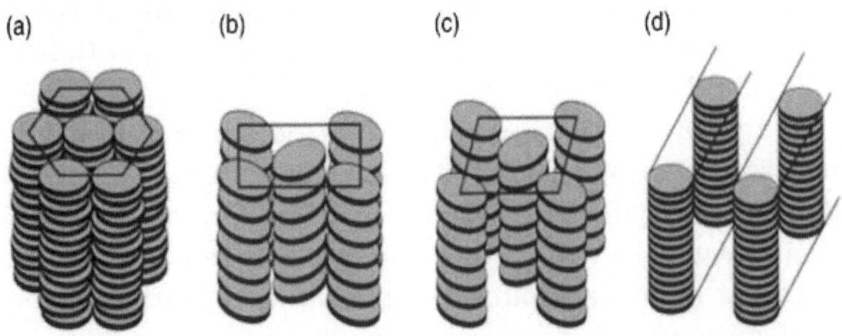

Figure 4. Columnar phase liquid crystals schematic diagram; (a) hexagonal, (b) rectangle, (c) oblique, and (d) lamellar [19].

Another classification of liquid crystal, lyotropic liquid crystal transitions, occurred by changing solvent concentration as well as varying the temperature. Amphiphilic molecules in a solution with hydrophobic and hydrophilic components resulted in diverse lyotropic liquid crystals and varied concentrations resulted in various circular or rod-shaped micelles in a solution. The rod-shaped micelles are arranged to form highly viscous hexagonal lyotropic liquid crystal while the circular micelle form a cubic pattern of lyotropic liquid crystal that does not have shear planes to make them more viscous than the hexagonal lyotropic phase. Shape-dependent numerous kinds of lyotropic liquid crystals can be obtained by manipulating the micelle shape present in the solution [20, 21].

3. Computational advancements

There have been relatively few computational approaches to the advancement of liquid crystals since their discovery. In the 1940s, Onsager showed the colloidal solution method for liquid crystal formation, and proved that transitions can be occurred from isotropic to anisotropic phase by changing the density [22]. Then, in the 1950s, Maier and Saupe developed the molecular mean-field theory that explained the temperature-dependent transitions of thermotropic liquid crystals [23]; however, in comparison to these theories, the first simulation theory linked to simple liquid crystals was developed in the 1990s [24]. In recent years, attempts have been made to investigate simulation approaches and methodologies for determining the functions and properties of liquid crystals. The most challenging aspect in molecular simulation is connecting different lengths as well as the time of action scales, and various models are employed to manage these simulations based on their time scales, as shown in Figure 5. Molecular bonds and intermolecular motions in the femto-second time scale range are addressed by quantum modeling, while atomistic modeling handles molecular bond motion to liquid crystal director alignment in the femto-to-nanosecond time scale. The coarse-grained approach facilitated simulations of liquid crystal polymers at nanosecond to microsecond time scales, whereas continuum modeling is better suited to simulations at longer length and time scales.

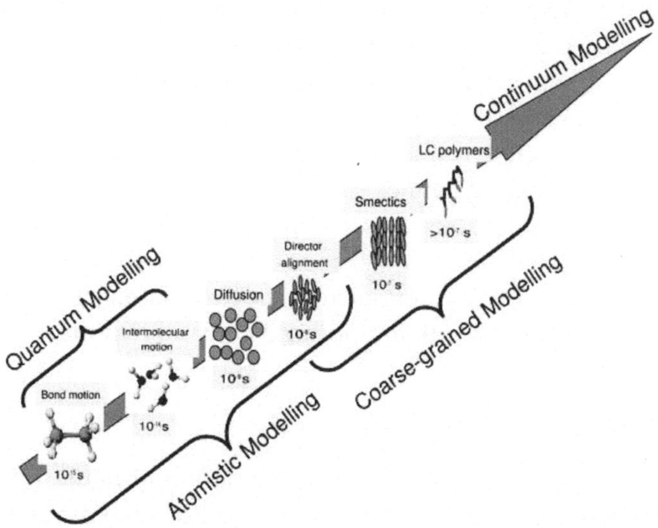

Figure 5. Computational schemes employed in liquid crystals research [25].

Quantum modeling mainly addressed the intermolecular motion with a time scale range of femto-seconds, whereas atomistic modeling is considered to be the vanguard of computational analysis of nanomaterials, and it plays a pivotal role as an interfacial region. The building complex models have also proven to be an effective tool for predicting the relationship between macroscopic properties and underlying atomistic structures. The structural analysis of a material is obtained at the end of molecular dynamics or Monte Carlo simulations, and it reflects the pair correlation function, which yields the probability of atomic separation by some calculated distance, and it can then be compared with experimental data to validate the computational as well as model methodology, ensuring that results correspond to the experimental setup. The coarse-grained models are vital for the illustration of complex systems that allow keeping the main entities and performing the important simulations that are not more suitable with other simulation methods like molecular dynamics having some limitations in time and length scales. The group of atoms is bounded by beads whose interaction depends on their internal energy function which accounts for the detailed simulations at the atomistic level for both bonded and non-bonded interactions. This modelling method can be applied to a wide range of length scales, such as a bead or interaction site of a group of atoms, a functional group, or a polymeric micro-particle, depending entirely on the system of interest and the investigated property [26].

Continuum models are proved as the most useful description of lateral composition segregation that measures the tendency of a solid solution to possess the phase separation. It has also been reported that continuum simulation is more feasible for more precise modeling of the mechanics problem of stress relaxation by lateral region formation of phase-separated material, where the system goes from constrained to less-stressed configuration and the tendency towards phase separation may decrease. The kinetics of morphological evolution reflects these transitions and is a superior alternative for simulations with larger lengths and time scales [27]. Molecular modeling becomes difficult as the length and time scales increase, thus various sophisticated simulation techniques are employed, including lattice Boltzmann nemato-dynamics [28], reduction of the Landau-de Gennes free energy [29], and continuum theory approaches [30]. The liquid crystal force field method is appropriate for a wide range of molecular stands. Bond stretching, bond bending, and torsional potential may be calculated using ab-initio / DFT [31] calculations and the force field approach in combination with sophisticated computer technology.

3.1 Recent classifications of liquid crystals

The recent advancements in computer simulation of liquid crystals classified them into three main categories; hard non-spherical models, soft non-spherical models and lattice models.

3.1.1 Hard non-spherical model

The hard non-spherical model is based on the fact that liquid crystal molecules are generally non-spherical hard ellipsoids, and hard sphero-cylinders, as shown in Figure 6. This intriguing feature would be useful in studying the basic effects and assessing the liquid crystal stability as well as other external stimuli interactions, whereas the phase behavior of these hard molecule-based models is reliant on density changes and is independent of temperature. This feature of liquid crystal molecules is demonstrated by Onsager's work [22], in which he deduced that transitions from isotropic to anisotropic occur well below the density at which the system would be expected to crystallize, provided that length is much greater than the diameter, and represent a non-spherical pattern. The interplay between translational and rotational entropy is the driving factor behind liquid crystal formation; as density increases the system minimizes its free energy by organizing itself, causing a rise in translational and rotational entropy.

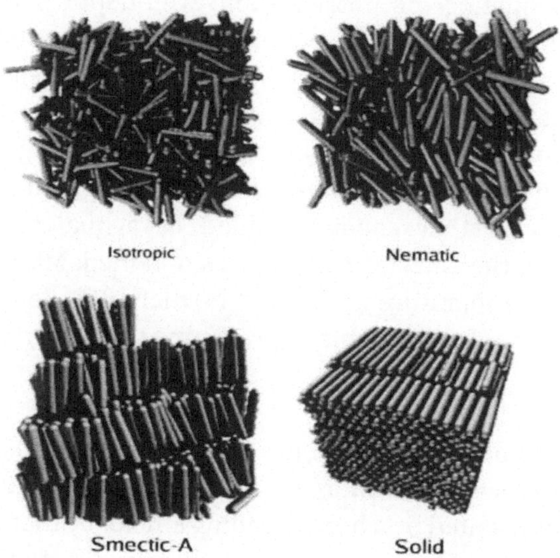

Figure 6. Simulations of hard sphero-cylinders with aspect ratio L/D = 5 at different densities [32].

3.1.2 Soft non-spherical model

Non-spherical simulations were done in the 1970s and subsequent years, but due to a lack of computer time, these simulations were not completed properly [33, 34, 35]. The original work has recently been expanded by combining thermodynamic and dynamical characteristics to calculate the value of electrons for non-spherical (ellipsoids) molecules and adding boundary constraints to decrease surface effects [36], revealing spontaneous liquid crystal ordering by an even compression of the isotropic state beyond the thermodynamic stable limit. However, using the molecular dynamic approach, it was discovered that the smectic phase develops from the nematic phase over time. This method allows for the comparison of diffusion coefficients parallel to smectic planes, which reveals mobility within layers and motion between layers in the perpendicular direction.

Liquid crystal molecules in a soft non-spherical model are soft and feature long-range molecular interactions, whereas temperature is used to determine thermodynamic equilibrium in this model, which was not studied in hard non-spherical models. The invention of the Gaussian overlap model [37] started work in this field in the 1970s, and the Gay-Berne potential is the most often used soft non-spherical model. This model, which was founded on the notion of fitting the potentials to a more exact atomic model of the molecule of interest [38], incorporates the attraction of soft ellipsoids, which have the same symmetry as hard ellipsoids. Although, the initial potential had unrealistic features, it has subsequently been improved and widely used [36]. The Gay-Berne (GB) model has the advantage of being able to explore a wide range of interactions between various thermotropic liquid crystals by adjusting the potential and well-depth parameters, as well as being simple to combine with other potentials. Recent results combine GB and a point dipole to illustrate the ferroelectric phase [39], as well as GB with a point quadrupole to demonstrate the smectic C phase [40]. Because this model includes a temperature factor, an isotropic phase was formed at high temperatures, but upon cooling, a nematic phase was observed, indicating that the smectic A phase was obtained, while further cooling yielded hexagonal alignment within the layers, indicating that the smectic B phase was obtained. Molecular diffusions can be identified as the temperature is reduced further [36], demonstrating that this advanced simulation model is capable of predicting thermodynamic transitions.

3.1.3 Lattice model

Lattice model has been popular in previous studies because of their simplicity and computational efficiency, such as the model shown in Figure 7, in which the chain stiffness is represented by a bond-bending efficiency that reduces the number of 90° lattice bends in some way. The most appealing feature of lattice model is that single-chain simulations may attain contour lengths of about 80,000 lattice spacing; nevertheless, the lack of small-angle bending deformations proposed by these models demonstrates the primary physical processes. Chains restricted by walls, for example, cannot show variations on the deflection length scale, but they can only show propagation along straight lines over lengths of the order of bond length (lb) or greater. Despite these significant limitations, these lattice models have some promising features, such as qualitatively describing the layering of effective monomers near hard walls, and large system sizes allowing a fairly accurate estimation of the phase diagram for capillary nematization for such liquid crystal polymers [42].

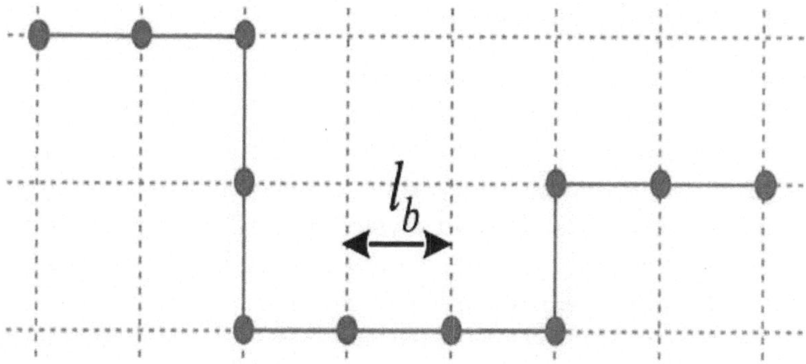

Figure 7. Schematic representation of self-avoiding walk on a simple cubic lattice for the simulation of polymers [41].

Simulation and theoretical reasoning have recently demonstrated the feasibility of extrapolating from high-dimensional systems to three-dimensional systems in the real world [25]. Even the most basic Lebwohl-Lasher lattice model has the correct phase transition properties, which could be used to support Maier and Saupe's original ideas that long-ranged, smoothly-varying interactions between 'swarms' of molecules are the most important factor in determining the stability of mesophases.

4. Applied applications

Liquid crystals have been considered peculiar substances since their discovery in 1888. The demand for low-power, high-efficiency displays invigorated research into liquid crystal electro-optical characteristics, and they offer a wide range of uses outside of displays, including switchable windows, thermometers, plasmonics, photovoltaics, and solar cells [43]. This section is devoted to exploring some useful and novel applications of LCs in different areas of day-to-day life as summarized in Table 1.

Table 1. Potential application domains of liquid crystal research.

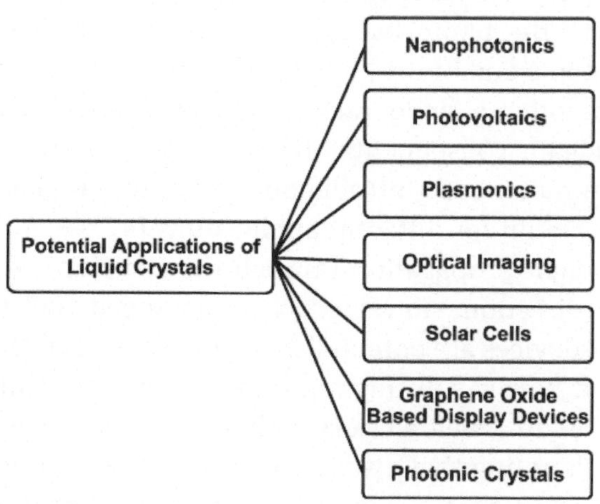

4.1 Graphene oxide liquid crystals

Graphene is an atomic-scale thin carbon material that has drawn much attention due to its unique and extraordinary electrical, chemical, thermal, mechanical and optoelectronic properties having a wide range of applications. Chemical oxidation of graphene produces graphene oxide, while wet chemical exfoliation of graphene produces graphene oxide-based liquid crystals [44]. Graphene oxide is unique in comparison to liquid crystal molecules because of its large diameter to thickness aspect ratio and optical anisotropy, exhibiting a strong magnetic response. Graphene oxide-based liquid crystals are employed in many display and device applications due to their electric field-induced birefringence. The devices based on graphene oxide consume less power than traditional liquid crystals and do not require specific electrode treatment. Also, the device size and switching time may be reduced by decreasing the size of the graphene oxide flakes [45]. When the size of the graphene oxide flakes decreases, the polarization anisotropy reduces quicker than the rotational viscosity, causing the rising time to increase, while falling time is controlled only by rotational viscosity, smaller flakes suffer less, resulting in a decrease in falling time, although, the smaller switching time is required for screen applications. Graphene oxide liquid crystals may also be utilized as rewritable boards with a dark or bright background [46], a reflective display that does not require polarizing optics and relies only on ambient light. These reflective displays which are controlled by an electric field are extensively used in electronic books because of their low energy consumption and inexpensive cost.

4.2 Photonic crystals

Liquid crystals have remarkable optical properties that respond to external stimuli such as light, temperature, electric field, magnetic field, and electrochemical reactions resulting in a dynamic change of material from the molecular to the macroscopic level, and this dynamic system makes them more useful in the nanofabrication of photonic materials. The responses of liquid crystals to various stimuli contribute to the photo bandgap tuneability, which has several uses in communication systems and optical integrated circuits. Photo crystals can control photons and hence offer a wide variety of uses based on defect engineering or the crystal bandgap; consequently, light can only be transmitted by adding defects, while the simultaneous introduction of point and line defects produces a miniature photonic circuit for different applications [47, 48, 49]. Photonic crystals may also influence light propagation through the structure resulting in a change in the propagation direction, slower propagation speed, and negative refraction. Controlled photonic devices are potential instruments for optical switching, routing, optical imaging, and power splitting because of their capabilities. This approach may also be employed to develop lasers with a narrow wavelength range on a tiny chip [50]. It is worth noting that, owing to better photonic crystal fabrication, 3D photonic crystals, rather than the currently utilized 2D crystals, will be used in future applications due to greater light control in the 3D range. Photonic crystal fibers can transmit a large range of light wavelengths with zero dispersion, and due to this uncommon property, they can be widely used for communication networks, polarization splitters, rotators, filters as well as multiplexers [43].

4.3 Photovoltaics

Liquid crystals are widely used in many photovoltaic devices owing to their unique intermolecular interactions. Discotic or columnar liquid crystals are more efficient to use and have more intermolecular interactions than inter-columnar interactions with special arrangements, in which each column acts as a molecular wire so that electrons and holes can move freely in one-dimensional structural arrangement and this alignment of molecules increases the electrical conductivity that allows them to be used in photovoltaic devices [51, 52]. Also, p-type and n-type discotic liquid crystals can be designed, and their properties, as well as alignment to the electrode surface, can be improved to achieve better charge migration results [53]. Calamitic or rod-like liquid crystals are also fascinating for semiconductor applications, particularly in display devices as well as conducting impurities [54]. In addition, liquid crystals create nanomorphology between the donor and acceptor layers to enhance film orientational order, resulting in efficient photocurrent generation and higher incoming photon conversion efficiency.

4.4 Nanophotonics

A newly emerging field that combines liquid crystals with nano-sized photonic structures is called nanophotonics [55], and is widely used in numerous applications. These devices are utilized to tune plasmonic nanostructures, as well as nanoporous materials, controlled magnetic field for switching liquid crystals and these liquid crystal composites with dielectric nanostructures are continuously gaining practical attention [56]. Liquid crystals are enabling to impart their long-ranged orientational symmetry to nanomaterials. Low concentrations of nanomaterials can be oriented as nematic patterns that can demonstrate improved physical properties of liquid crystals as well as a strong response to external stimuli such as electro-optic and magneto-optic responses, electrochemical memory effect, and changes in liquid crystal orientational order parameters [57]. Furthermore, the ability of liquid crystals to flow and fill nano-gaps led to the development of adjustable photonic crystal devices, porous silicon thermal tuning, and 2D photonic bandgap structures filled with liquid crystals [58]. Because nematic liquid crystals have poor magnetic anisotropy, switching the device requires a high magnetic field of about one Tesla, which would damage the device, so combining liquid crystals with ferromagnetic nanoparticles can be used to improve device switching at low magnetic fields for even better practical applications [59] that would result in some intriguing effects like increased dielectric anisotropy, increased order parameter as manifested by an increase in birefringence, enhanced non-linear effects, and so on, and these factors resulted in a decrease in threshold voltage, which is an important factor of power consumption in display applications [60].

4.5 Plasmonics

Liquid crystals have gained much attention for plasmonic applications because of their attractive features such as; large birefringence, less driving threshold and various possible fabrication methods. This particular application of liquid crystals opens up the opportunities for next-generation plasmonic devices by combining both the liquid crystals and nanostructures with enhanced plasmonic properties resulted in devices such as; modulators, absorbers, plasmonic waveguides, plasmonic switches and color filters, etc. The main feature of liquid crystals that makes them suitable for active plasmonic devices is the difference in refractive index resulting in large birefringence. There are numerous ways to apply liquid crystals to the plasmonic nanostructures to get a promising change in refractive index that is essential for its potential device applications. Liquid crystal-based active plasmonic devices can be categorized according to the driving methods such as; electric field-driven method, heat-driven method, acoustic waves driven method and light-driven method. Depending on whether external biasing is supplied, such as an electric field, heat, light, or acoustic waves, liquid crystal molecules will align properly when exposed to external stimuli. This feature of liquid crystals is coupled with a periodic nanostructured material to produce the consequent controlled effect of plasmonic structures such as reflection, transmission, or absorption, indicating its potential in a variety of plasmonic-based devices [61]. Hence the combination of liquid crystals and metallic nanoparticles can be utilized to create tunable plasmonic devices, because metallic nanoparticles exhibit strong attenuation at resonance wavelengths that correspond to localized surface plasmon and this resonance wavelength can be modified by varying surrounding refractive index which can be controlled by external electric field using liquid crystals, yielding promising results [62, 63].

5. Current challenges and problems

Since the previous decade, liquid crystals have been a popular issue, with considerable progress made in understanding their complicated process and alignments in many applications, particularly in display devices, which makes them fascinating. However, quite some important alignment aspects are not understood well specifically regarding liquid crystals molecular interactions, which is the biggest challenge in the liquid crystals industry of this era. There is a need for the developed alignment technology that would resolve the existing limiting factors that are affecting display quality and processing costs. The specific challenging areas of liquid crystals include non-uniform alignment over large displays, image sticking, uneven display brightness and multi-domain alignment pattern configurations [64]. Response time of liquid crystals is another challenge that plays an important role in display applications. Slow response time of liquid crystal causes blur quality and undesirable image display which is due to the non-adjustable pre-tilt angle of liquid crystals with the substrate that eventually affects the cell dynamics. To improve image quality and display, a proper oriented angle adjustment between liquid crystals and substrate is needed to be addressed [65]. Furthermore, anchoring energy, which is the energy necessary to deviate the director off the molecular axis at a specific angle, may be used to evaluate the alignment strength of the liquid crystal. Associated with the pre-tilt angle, stability of anchoring energy is another biggest challenge of liquid crystals industry to obtain uniform planer alignment, and to overcome this problem, rubbing mechanism was introduced to align liquid crystals, but this mechanism also faces challenges including static charge accumulation as well as dust-particle generation on crystal surface which affects the basic device characteristics and operational mechanism. The image sticking phenomena that happens when an image is shown for a long time is an essential aspect in determining the liquid crystal display quality. The present state of the problem of the liquid crystal [64] is indeed the electric charge or residual DC charge produced on the liquid crystal surface during long-term picture display, which causes selective surface adsorption of ionic impurities present in liquid crystal material or layers. Moreover, liquid crystals have some other drawbacks related to a huge number of ionic impurities present in aligned layers of liquid crystals and the presence of conjugated functional groups or lone pairs in liquid crystal molecules and between its aligned layers that would eventually result in disturbing as well as disrupting the display application, hence affecting the role of liquid crystals in device applications [66].

6. Future prospects

Liquid crystals are materials with extraordinary properties mainly because of the partial orientational order having application in electronic, display and non-display devices that have become a pervasive as well as unavoidable feature in routine life. The full exploration of these materials is still a challenge due to the tricky problems involved in its phase structures, and a wide range of thermotropic liquid crystals with different shapes are reported with molecular self-assembly and interactions. Despite molecular shape and properties, fascinating research trends highlights the lack of systematic studies. However, this gap can be considered for future research by developing standard techniques and procedures. In addition, the molecules can be modified by the functional units to obtain highly functional systems using a variety of unexplored solutions.

The progression in nanophotonics can be used in combination with liquid crystals to design tunable devices with the help of powerful computer simulations which would continue assisting the building and understanding of liquid crystal devices at atomistic as well as a molecular level. Because of developing nano-sized liquid crystals, the future decade is likely to reveal more novel adjustable technologies, and this area needs strong developments to generate models as well as techniques with enhanced phase behavior representation diagrams which would lead to study the structural changes along-with phase behavior to explore novel material properties. Polyphilic molecules, which provide various interaction regions in a molecule to incorporate other flexible materials, are another intriguing feature of liquid crystals, and they can be more beneficial by controlling the molecular self-assembly, which is the most difficult part of synthetic engineering. However, if simulations are sophisticated enough to be utilized as engineering tool to create polyphilic liquid crystals with the desired structures; the liquid crystal industry can achieve amazing success.

7. Concluding remarks and summary

This chapter summarized an overview of liquid crystals by exploring fundamentals along with its recent computational advancement, challenges faced, applied applications and future prospects. Liquid crystals are unique in their properties having potential for numerous applied applications ranging from optical imaging, plasmonics, solar cells, photonic crystals, photovoltaics as well as nanophotonics, and the research in this field is the spark for the next forefront of economic development. Because of the combination of LCs with nanotechnology, the next decade will see noticeable developments and an increase in miniaturized tunable devices such as; tunable optical filters based on LCs and plasmonic phenomena like enhanced optical transmission through nano-slits, extinction of metallic nanoparticles and nanostructures, and LC biosensors that make use of LC and nanostructure characteristics.

References

1. Lagerwall JP, Scalia G. A new era for liquid crystal research: Applications of liquid crystals in soft matter nano-, bio-and microtechnology. Current Applied Physics. 2012;12(6):1387-1412
2. Körner H, Shiota A, Bunning TJ, Ober CK. Orientation-on-demand thin films: Curing of liquid crystalline networks in ac electric fields. Science. 1996;272(5259):252-255
3. Hulvat JF, Stupp SI. Liquid-crystal templating of conducting polymers. Angewandte Chemie International Edition. 2003;42(7):778-781
4. Demus D, Goodby JW, Gray GW, Spiess HW, Vill V, editors. Handbook of Liquid Crystals, Volume 2A: Low Molecular Weight Liquid Crystals I: Calamitic Liquid Crystals. USA: John Wiley & Sons; 2011
5. Goodby JW, Collings PJ, Kato T, Tschierske C. In: Gleeson H, Raynes P, Vill V, editors. Handbook of Liquid Crystals, 8 Volume Set. Vol. 1. USA: John Wiley & Sons; 2014
6. Frank FC. I. Liquid crystals. On the theory of liquid crystals. Discussions of the Faraday Society. 1958;25:19-28
7. Sood S & Raina KK. Induced Alignment in Nematic Liquid Crystal by Dichro Disperse Orange 3 Azo dyeic [Doctoral dissertation]. 2013

8. Sluckin T, Dunmur D, Stegemeyer H. Crystals That Flow. London: Taylor & Francis; 2004. pp. 381-385
9. Tamhane K. Formation of Lyotropic Liquid Crystals Through the Self-Assembly of Bile Acid Building Blocks [Electronic Theses and Dissertations]. USA: Stars Citation, University of Central Florida; 2009
10. Stark H. Physics of colloidal dispersions in nematic liquid crystals. Physics Reports. 2001;351(6):387-474
11. Pavlin J, Vaupotič N, Čepič M. Liquid crystals: A new topic in physics for undergraduates. European Journal of Physics. 2013;34(3):745
12. Carlescu I. Introductory chapter: Liquid crystals. In: Liquid Crystals-Self Organized Soft Functional Materials for Advanced Applications. Rijeka: IntechOpen; 2018
13. Friedel G. Mesomorphic states of matter. Annales de Physique. 1922;9(18):273-474
14. Matsui T. Numerical simulation of lasing dynamics in choresteric liquid crystal based on ADE-FDTD method. In: Numerical Simulations of Physical and Engineering Processes. UK: IntechOpen; 2011. p. 207
15. Gray GW. The mesomorphic behaviour of the fatty esters of cholesterol. Journal of the Chemical Society (Resumed). 1956;726:3733-3739
16. Oswald P, Pieranski P. Nematic and Cholesteric Liquid Crystals: Concepts and Physical Properties Illustrated by Experiments. USA: CRC Press; 2005
17. Sluckin T, Dunmur D, Stegemeyer H. Crystals That Flow. London: USA: Taylor & Francis; 2004. pp. 381-385
18. Kumar S. Self-organization of disc-like molecules: Chemical aspects. Chemical Society Reviews. 2006;35(1):83-109
19. Oswald P & Pieranski P. Smectic and Columnar Liquid Crystals: Concepts and Physical Properties Illustrated by Experiments. USA: Taylor & Francis; 2005
20. Liu Q, Tang J, Zhang Y, Martinez A, Wang S, He S, et al. Shape-dependent dispersion and alignment of nonaggregating plasmonic gold nanoparticles in lyotropic and thermotropic liquid crystals. Physical Review E. 2014;89(5):052505. USA: American Physical Society
21. Collings PJ. Phase Structures and Transitions in Thermotropic Liquid Crystals. In: Handbook Of Liquid Crystal Research. USA: Swartmore College; 1997. pp. 99-124
22. Onsager L. The effects of shape on the interaction of colloidal particles. Annals of the New York Academy of Sciences. 1949;51(4):627-659
23. Maier W, Saupe A. A simple molecular theory of the nematic crystalline liquid state. Zeitschrift für Naturforschung A. 1958;13(7):564-566
24. Neal MP, Parker AJ. Computer simulations using a longitudinal quadrupolar Gay–Berne model: Effect of the quadrupole magnitude on the formation of the smectic phase. Chemical Physics Letters. USA: Swartmore College; 1998;294(4-5):277-284

25. Wilson MR. Progress in computer simulations of liquid crystals. International Reviews in Physical Chemistry. 2005;24(3-4):421-455

26. Casalini T. Fundamentals and application of modeling in support of spinal cord injury repair strategies. In: Spinal Cord Injury (SCI) Repair Strategies. Netherlands: Elsevier; 2020. pp. 279-306

27. Johnson HT. Effects of stress on formation and properties of semiconductor nanostructures. In: Material Substructures in Complex Bodies. Netherlands: Elsevier; 2007. pp. 284-313

28. Care CM, Halliday I, Good K, Lishchuk SV. Generalized lattice Boltzmann algorithm for the flow of a nematic liquid crystal with variable order parameter. Physical Review E. 2003;67(6):061703

29. Dupuis A, Marenduzzo D, Yeomans JM. Numerical calculations of the phase diagram of cubic blue phases in cholesteric liquid crystals. Physical Review E. 2005;71(1):011703

30. Davies JB, Day S, Di Pasquale F, Fernandez FA. Finite-element modelling in 2-D of nematic liquid crystal structures. Electronics Letters. 1996;32(6):582-583

31. Iqbal MA, Ashraf N, Shahid W, Afzal D, Idrees F, Ahmad R. Fundamentals of Density Functional Theory: Recent Developments, Challenges and Future Horizons [Online First]. IntechOpen; 2021. Available from: https://www.intechopen.com/online-first/78124. DOI: 10.5772/intechopen.99019

32. McGrother SC, Williamson DC, Jackson G. A re-examination of the phase diagram of hard spherocylinders. The Journal of Chemical Physics. 1996;104(17):6755-6771

33. Vieillard-Baron J. The equation of state of a system of hard spherocylinders. Molecular Physics. 1974;28(3):809-818

34. Baumgärtner A, Binder K, Hansen JP, Kalos MH, Kehr K, Landau DP, et al. Applications of the Monte Carlo Method in Statistical Physics. Vol. 36. Germany: Springer Science & Business Media; 2013

35. Boublík M, Pivec L, Šponar J, Šormová Z. Deoxyribonucleic acids in solution. V. Heat denaturation of fractions and subunit structure of calf thymus DNA. Collection of Czechoslovak Chemical Communications. 1965;30(8):2645-2653

36. Allen MP, Wilson MR. Computer simulation of liquid crystals. Journal of Computer-Aided Molecular Design. 1989;3(4):335-353

37. Berne BJ, Pechukas P. Gaussian model potentials for molecular interactions. The Journal of Chemical Physics. 1972;56(8):4213-4216

38. Gay JG, Berne BJ. Modification of the overlap potential to mimic a linear site–site potential. The Journal of Chemical Physics. 1981;74(6):3316-3319

39. Berardi R, Orlandi S, Zannoni C. Antiphase structures in polar smectic liquid crystals and their molecular origin. Chemical Physics Letters. 1996;261(3):357-362

40. Neal MP, Parker AJ. Computer simulations using a longitudinal quadrupolar Gay–Berne model: Effect of the quadrupole magnitude on the formation of the smectic phase. Chemical Physics Letters. 1998;294(4-5):277-284

41. Binder K, Egorov SA, Milchev A, Nikoubashman A. Understanding the properties of liquid-crystalline polymers by computational modeling. Journal of Physics: Materials. 2020;3(3):032008

42. Ivanov VA, Rodionova AS, An EA, Martemyanova JA, Stukan MR, Müller M, et al. Orientational ordering transitions of semiflexible polymers in thin films: A Monte Carlo simulation. Physical Review E. 2011;84(4):041810

43. Obayya S, Hameed MFO, Areed NF. Computational Liquid Crystal Photonics: Fundamentals, Modelling and Applications. UK: John Wiley & Sons; 2016

44. Zheng Q, Li Z, Yang J, Kim JK. Graphene oxide-based transparent conductive films. Progress in Materials Science. 2014;64:200-247

45. Ahmad RTM, Shen TZ, Masud AR, Ekanayaka TK, Lee B, Song JK. Guided electro-optical switching of small graphene oxide particles by larger ones in aqueous dispersion. Langmuir. 2016;32(50):13458-13463

46. He L, Ye J, Shuai M, Zhu Z, Zhou X, Wang Y, et al. Graphene oxide liquid crystals for reflective displays without polarizing optics. Nanoscale. 2015;7(5):1616-1622

47. Akahane Y, Asano T, Song BS, Noda S. Investigation of high-Q channel drop filters using donor-type defects in two-dimensional photonic crystal slabs. Applied Physics Letters. 2003;83(8):1512-1514

48. Takano H, Song BS, Asano T, Noda S. Highly efficient multi-channel drop filter in a two-dimensional hetero photonic crystal. Optics Express. 2006;14(8):3491-3496

49. Okano M, Kako S, Noda S. Coupling between a point-defect cavity and a line-defect waveguide in three-dimensional photonic crystal. Physical Review B. 2003;68(23):235110

50. Takano H, Akahane Y, Asano T, Noda S. In-plane-type channel drop filter in a two-dimensional photonic crystal slab. Applied Physics Letters. 2004;84(13):2226-2228

51. Iqbal MA, Malik M, Shahid W, Din SZ, Anwar N, Ikram M, et al. Materials for Photovoltaics: Overview, Generations, Recent Advancements and Future Prospects [Online First]. UK: IntechOpen; 2022. Available from: https://www.intechopen.com/online-first/80148. DOI: 10.5772/intechopen.101449

52. Balagurusamy VSK, Prasad SK, Chandrasekhar S, Kumar S, Manickam M, Yelamaggad CV. Quasi-one dimensional electrical conductivity and thermoelectric power studies on a discotic liquid crystal. Pramana. 1999;53(1):3-11

53. Andrienko D, Feng X, Marcon V, Pisula W, Kirkpatrick J, Grozema F, et al. Rational design of the shape and periphery of discotics: A synthetic way towards high charge carrier mobilities. In: APS March Meeting Abstracts. USA: APS (American Physical Society); 2009. pp. Y36-Y011

54. AlKhalifah MS, Lei C, Myers SA, O'Neill M, Kitney SP, Kelly SM. Solution-processed bilayer photovoltaic devices with nematic liquid crystals. Liquid Crystals. 2014;41(3):402-417

55. Iqbal MA, Ashraf N, Shahid W, Awais M, Durrani AK, Shahzad K, et al. Nanophotonics: Fundamentals, Challenges, Future Prospects and Applied Applications. UK: IntechOpen; 2021

56. Abbate G, Vita F, Marino A, Tkachenko V, Slussarenko S, Sakhno O, et al. New generation of holographic gratings based on polymer-LC composites: POLICRYPS and POLIPHEM. Molecular Crystals and Liquid Crystals. 2006;453(1):1-13

57. Dierking I, Scalia G, Morales P. Liquid crystal–carbon nanotube dispersions. Journal of Applied Physics. 2005;97(4):044309

58. Weiss SM, Ouyang H, Zhang J, Fauchet PM. Electrical and thermal modulation of silicon photonic bandgap microcavities containing liquid crystals. Optics Express. 2005;13(4):1090-1097

59. Zadorozhnii VI, Vasilev AN, Reshetnyak VY, Thomas KS, Sluckin TJ. Nematic director response in ferronematic cells. EPL (Europhysics Letters). 2005;73(3):408

60. Li F, West J, Glushchenko A, Cheon CI, Reznikov Y. Ferroelectric nanoparticle/liquid-crystal colloids for display applications. Journal of the Society for Information Display. 2006;14(6):523-527

61. Si G, Zhao Y, Leong ESP, Liu YJ. Liquid-crystal-enabled active plasmonics: A review. Materials. 2014;7(2):1296-1317

62. Hsu LH, Lo KY, Huang SA, Huang CY, Yang CS. Irreversible redshift of transmission spectrum of gold nanoparticles doped in liquid crystals. Applied Physics Letters. 2008;92(18):181112

63. Iqbal MA, Malik M, Shahid W, Ahmad W, Min-Dianey KA, Pham PV. Plasmonic 2D Materials: Overview, Advancements, Future Prospects and Functional Applications. UK: IntechOpen; 2021. Available from: https://www.intechopen.com/online-first/79872

64. Ishihara S. How far has the molecular alignment of liquid crystals been elucidated? Journal of Display Technology. 2005;1(1):30

65. Scheffer TJ, Nehring J. Accurate determination of liquid-crystal tilt bias angles. Journal of Applied Physics. 1977;48(5):1783-1792

66. Wang Z, Cai S. Recent progress in dynamic covalent chemistries for liquid crystal elastomers. Journal of Materials Chemistry B. 2020;8(31):6610-6623. UK: Royal Society of Chemistry

3

Phase Transitions and Structure of Liquid Crystalline Cellulose Ether Solutions in a Magnetic Field and in Its Absence

1. Introduction

1.1 Liquid crystalline state of matter

Matter can exist in the following phase states: crystalline, amorphous (liquid), gaseous, and liquid crystalline. The substances in the different phase states exhibit the different spatial arrangements of structural elements (molecules, atoms, ions).

The crystalline phase state is characterized by a three-dimensional long-range order in arrangement of atoms or molecules. Long-range orientational order is an order observed over distances hundreds and thousands of times larger than the size of the molecules and can exist in one, two, or three dimensions.

The amorphous phase state (liquid phase state) is characterized by a short-range orientational order, i.e., an order observed over distances commensurate with molecular size. In the vicinity of any given molecule, its neighbors can be ordered but at a small distance from the molecule. The amorphous phase state is attributed to all liquids (excluding mercury) as well as glass and other solid noncrystalline substances. In a gaseous phase state, there is no order in the spatial arrangement of structural elements.

The term "liquid crystal" is a contradiction in itself. The crystals are considered to be anisotropic solids showing usually very low deformations even under high mechanical stress. On the other hand, the fluids are able to flow, and they exhibit isotropic optical properties. Meanwhile, over a hundred years ago, it was found [1] that some organic substances in melt, that is they are able to flow, exhibit anisotropic optical properties (birefringence). Such a substance was first described in 1888 by F. Reinitzer [2], Professor of the Botanic School of Graz University of Technology (Austria). He examined the physiological activity of cholesterol, the chemical formula for which had not yet been determined. In the course of his studies, F. Reinitzer synthesized cholesterol derivatives and determined their melting temperatures. He found that cholesteryl benzoate under melting at 145.5°C turned into a cloudy liquid and after further heating, turned into transparent melt at 178.5°C. The refractive indices of the cloudy melt measured in two perpendicular directions were different, which proved its anisotropic properties. Therefore, F. Reinitzer assumed that the turbid melt included two substances. F. Reinitzer tried to separate the substances, but he could not, and so he sent a cholesteryl benzoate sample for examination to physicist, Prof. Lehmann, who lived in Dresden. Prof. Lehmann examined the sample with a polarizing microscope and concluded that the turbid melt did not comprise two different liquids but presented a single liquid crystal [3]. Thus, it was Lehmann who introduced the term "liquid crystal." In 1890, he discovered liquid crystals of ammonium oleate, para-azoxyanisole, and para-azoxyphenetole.

The nature of a liquid crystalline (LC) phase state is as follows: some substances, as soon as they reach the temperature at which the three-dimensional lattice disintegrates, do not exhibit the direct transition to isotropic liquid but hold intermolecular order. In LC systems the order is not three-dimensional but, according to Gray, one- or two-dimensional, i.e., the regularity is partially destroyed, but the long-range orientational order always remains in one or two directions. Such behavior determines, on the one hand, an essential fluidity (ability for irreversible deformation) and, on the other hand, a demonstration of anisotropic physical properties in contrast to liquids possessing a short-range order.

Further detailed study of LC substances revealed that the crystal-liquid crystalline transition and consequent liquid crystalline-liquid (amorphous) transition are phase transitions of the first order and that the liquid crystals are in a special phase state, which is neither conventional crystalline nor amorphous phase state.

Liquid crystalline state is a thermodynamically stable phase state at which the matter exhibits the stable anisotropic physical properties of crystalline solids and the fluidity typical for liquids [4].

As liquid crystal exists between crystalline and amorphous phase states, it is called "mesomorphic phase" ("mesos" in Greek means middle, intermediate). The term "mesomorphic phase" or "mesophase" was introduced by Friedel in 1922.

The majority of liquid crystals have an enantiotropic mesophase, i.e., they appear upon both the melting of a crystalline solid and the cooling of a melt. However, some liquid crystals possess a monotropic mesophase appearing only upon cooling [5].

The main cause of matter transition to a liquid crystalline phase state, as soon as the melting point is achieved, is the asymmetric molecule structure. All matters demonstrating the transition to mesophase consist of the elongated molecules. At Tmelt, the intermolecular forces of lattice are not enough to hold molecules in the fixed position, and the three-dimensional lattice breaks down. On the other hand, the molecule anisotropy is an important factor of the preservation of a certain degree of intermolecular order. An additional kinetic energy is required to break this relative order too. So, a so-called step-by-step transition is observed from true crystal order to a random amorphous phase state. Each of these "steps" is characterized by strictly defined latent heat of transition. As a rule, the latent heat of the liquid crystalline-amorphous state transition is not high [5]. The basic change occurs at the initial disintegration of the three-dimensional crystalline lattice.

As mentioned above, the liquid crystalline systems demonstrate one- and two-dimensional orders [1]. A one-dimensional order of molecules means only a long-range orientation order along the molecule axes (Figure 1).

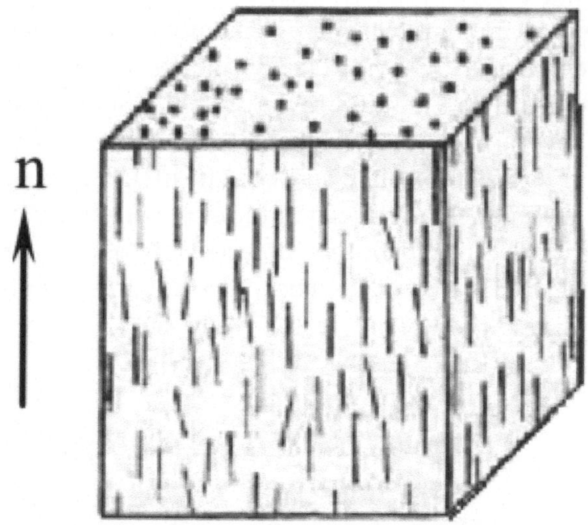

Figure 1. Nematic mesophase [5].

The gravity center of individual molecules is not coordinated relative to each other, and molecules can have a casual azimuthal deviation from the main axis. Such a structure is called nematic. Nematic systems are the first type of liquid crystal. For example, п-asocsianisol (PAA) [5]:

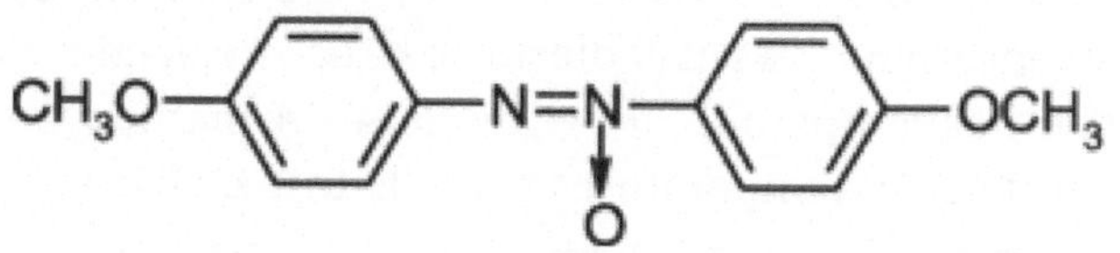

The patterns of so-called cholesteric systems are more ordered in comparison to nematic liquid crystals (Figure 2).

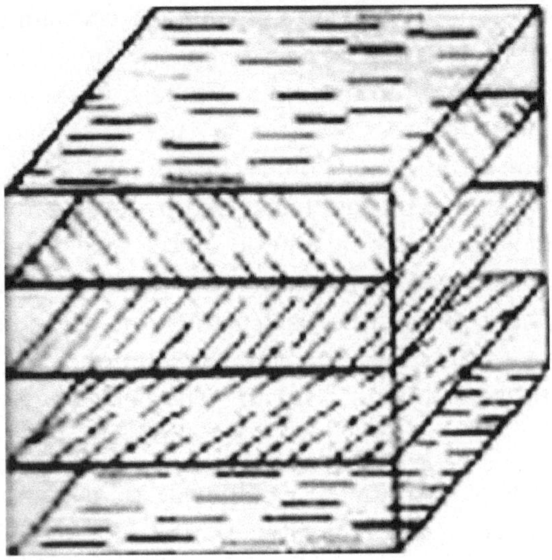

Figure 2. Cholesteric mesophase [5].

Their structure is a combination of parallel nematic layers where the direction of axes of molecules within each following layer is turned on a certain angle concerning the direction of molecule axes in the previous layer. Thus, a helix appears with a pitch that can be of several hundreds of nanometers. The equidistant arrangement of parallel layers and constant helix pitch allow the formal classification of this type of structure as two-dimensional; however, the molecule orientation within the layers is of a nematic nature due to which the cholesteric liquid crystals are sometimes referred to as a variant of the nematic phase. The cholesteric liquid crystals are so called because they are the complex esters of cholesterol, mainly:

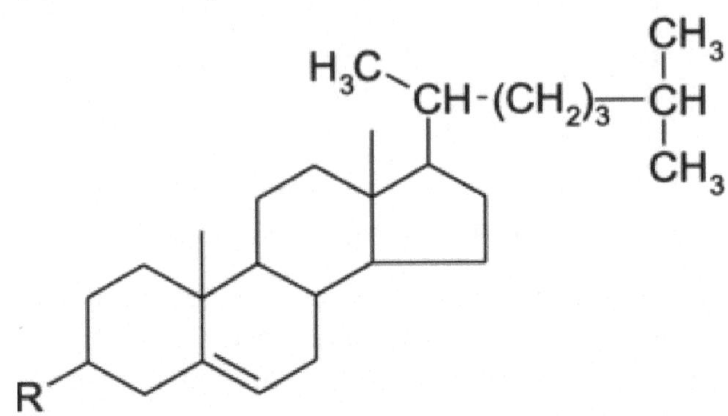

where R is the residue of relevant acid. But the complex esters of cholesterol are not the only representatives of cholesteric structure. A cholesteric mesophase is characteristic to other compounds, for example, 4′-(4-methoxybenzyl idenamin)aril cinnamate over a range of temperatures from 82 to 102°C [5]:

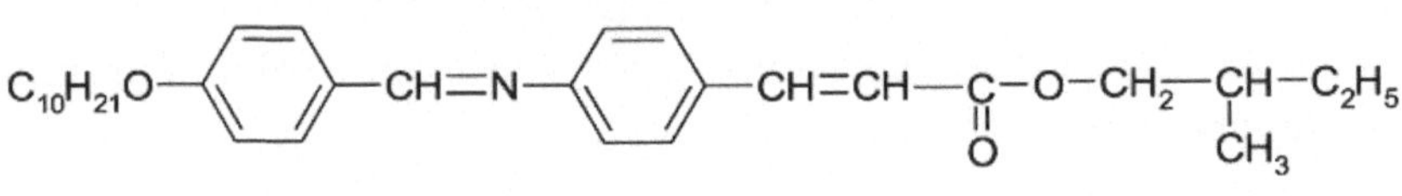

All such compounds contain asymmetric (chiral) atoms of carbon. Chiral properties give rise to the formation of a cholesteric mesophase. If the mesogenic compound is able to form a nematic phase and its molecules are chiral, then a cholesteric mesophase appears. This is proved by the fact that an addition to the nematic of a small amount of the cholesteric or chiral compound, which is not mesogenic, transfers the latter to the cholesteric. It should be noted that when changing the external conditions, the helix pitch may alter by discrete quantities up to infinity. Under such conditions, the cholesteric liquid crystal-nematic liquid crystalline transition is observed.

The smectic liquid crystals are the most ordered. They represent two-dimensional liquid crystals: the centers of molecule mass are located in layers, but the director n of each layer is no longer aligned along the layer surface but is tilted at a certain angle to it (Figure 3).

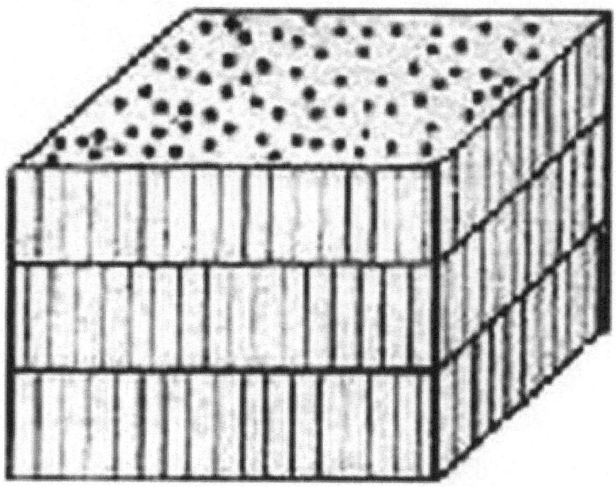

Figure 3. Smectic A mesophase [5].

These liquid crystals are usually referred to smectic A, which is the most commonly found among smectic liquid crystals. The typical example of A smectic is ethyl ester of n-nitrooxybenzoic acid (ENB) [5]:

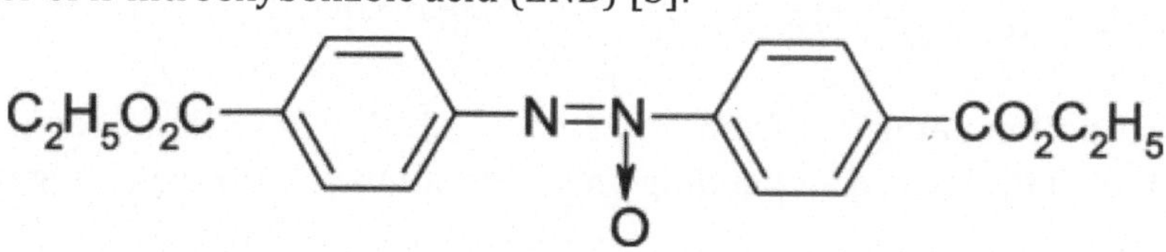

within temperatures from 114 to 120°C.

The thickness of layers is usually equal to the length of molecules of mesogen and the majority of the smectics exhibits such property. But for some smectics, the layer thickness is less than the molecular length or exceeds it. This is determined by the peculiarities of molecular structure.

For example, the layer thickness of 4-cyano-4′-n-octilbiphenyl:

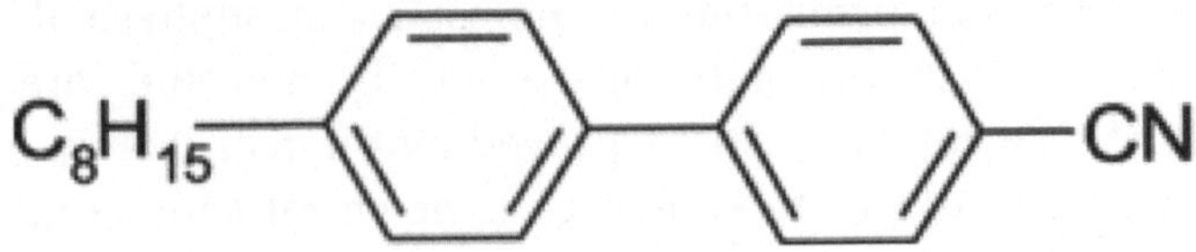

exceeds the molecule length by 1.4 times. In this case, a double-layered structure seems to exist and as such molecules within one layer enter the gaps between the molecules of the other layer.

The smectics with non-ordered layers are called smectic C or tilted smectics (Figure 4).

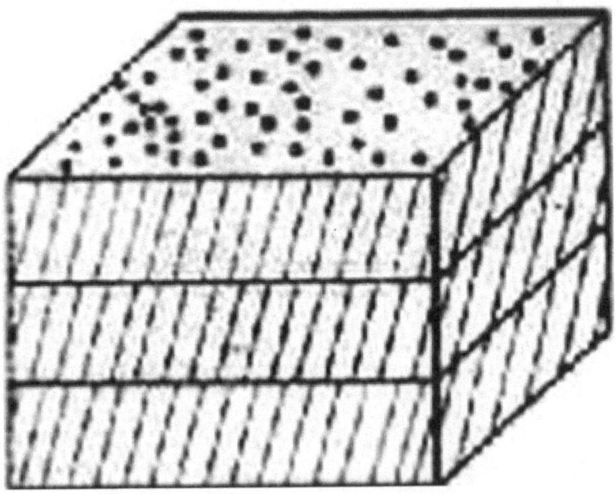

Figure 4. Smectic C mesophas.

They are commonly classified into smectic C with a small ($\omega < 30°$) and big ($\omega \approx 45°$) tilt angle of the director to the normal layer. Upon heating the smectic C with a small tilt angle transfers into smectic A. However, some compounds, such as terephtal-bis-4-n-butylaniline (TBBA):

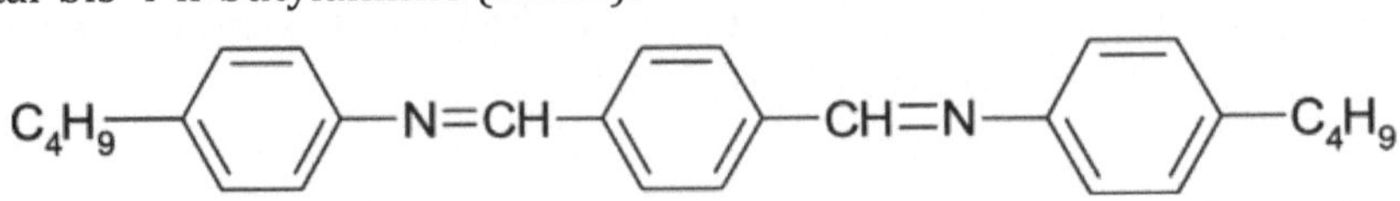

within the temperatures from 114.1 to 172.5°C, where the smectic C phase exists, exhibit a tilt angle ω depending on temperature and upon heating tilt angles become zero.

In such structures, the director processes around the cone from one layer to the next. Such a structure is similar to cholesteric.

Liquid crystals appearing upon cooling or heating by individual matters are called thermotropic. Liquid crystals resulting from the dissolution of matters are called lyotropic.

In [4], the mesogenic compounds are classified into the following groups based on their chemical structure:

- Aromatic compounds without bridge groups;
- Heteroaromatic compounds without bridge groups;
- Aromatic compounds with one bridge group;
- Aromatic compounds with several similar bridge groups;
- Aromatic compounds with different bridge groups;
- Stilbene, amides of carbon acid, derivatives of hydrazine and glyoxal;
- Aromatic carbon acids;
- Carbonic acid salts and ammonium salts;
- Alicyclic and aliphatic compounds.

Following the above classification, the various chemical fragments used in synthesis of mesogenic compounds are listed below (R is the n-alkyd; R' is the branched or unsaturated alkyl): most often used:

—OR, —R, —COOR, —OOCR, —OOCOR, '—CH=CH—, COOR.

often used:

$$—CN, —COR, —NO_2, —Cl.$$

seldom used:

—H, —F, —BR, —I, —R', —N=C=O, —OH, —OR', —OCOR', —COOR',
—CR=CR—COOR, $—NH_2$, —NHR, $—NR_2$, —NHCHO, —NHCOR, —SR,
—COSR, —OCOSR, —HgCl, $—HgOCOCH_3$, $—OCF_3$, $—O(CH_2)_nOR$, $—CH_2CH_2OH$

If one and the same substance has both nematic and smectic phases, the temperature of a smectic phase is always lower than the temperature of a nematic phase. Upon heating or cooling of the matter with the symmetrical molecules, the phase transitions from crystalline solids (CS) to isotropic liquids (IL) occur according to the scheme:

$$CS \leftrightarrow SLC \leftrightarrow NLC \leftrightarrow IL.$$

Transition temperatures are repeatable and reversible in case of enantiotropic transitions. Matters with optically active molecules exhibit the phase transitions according to the following scheme:

$$CS \leftrightarrow SLC \leftrightarrow ChLC \leftrightarrow IL.$$

1.2 Specific behavior of polymer liquid crystalline state

The specifics of polymers include both nonequilibrium states and manifestations of features, typical of an LC state, when the polymer is in an amorphous but not in a mesomorphous state. The majority of flexible chain polymers are subjected to orientational stretching during their manufacturing. If this is followed by a glass transition of the polymer system, the orientational order remains practically infinitely long. The resulting structure is nonequilibrium since after prolonged heating and slow cooling down to the original temperature, the system exhibits a clear tendency to disorder through weakening anisotropic properties [1].

The difference in phase transitions between the low-molecular matters and polymers should be assumed not in the special thermodynamic behavior of mesophase but rather in the kinetics of these transformations. The higher speed of phase transitions in low-molecular systems is quite clear as well as the lower speed of such transformations in polymers. These general properties of polymer systems and transitions to LC state and especially the liquid crystalline-crystalline solid transition, such as direct amorphous-crystalline solid transition, can be accompanied by such long periods of induction that they are not comparable to internal observation.

It should be noted that the rigid-chain polymers that most probably form the liquid crystals due to the high geometric anisotropy of molecules melt at temperatures higher than the temperatures of their intensive thermal destruction. Thus, polymers (with few exceptions) are unlikely to produce thermotropic LC systems. Apparently, this allowed Ph.H. Geil to consider the formation of only lyotropic liquid crystals though at present, thermotropic high-molecular LC systems are also known; for example, the melt of polypropylene, hydroxypropyl cellulose, as well as system based on block copolymers [5, 6, 7, 8].

The appearance of a mesophase in polymer systems is caused by the following [1]:

1. The ordering caused by the interaction of rather long side groups (side chains) in polymers.
2. The ordering caused by the interaction of homogeneous sequences (blocks) in block copolymers.
3. The ordering caused by the molecule rigidity. The distribution of rigid macromolecules in melts and solutions cannot be random. The polymer systems with rigid macromolecules exhibit a spontaneous transition to ordered states, which do not achieve a three-dimensional order; they are limited bý the one- or two-dimensional order.

The degree of orientational order is mainly determined by the flexibility of macromolecules, which is measured by Khun's segment A. The values of Khun's segments and the parameters of the intermolecular orientational order Q of different polymers are given in Table 1.

Table 1. Values of Q and A for polymers of various molecular weights [6].

Polymer	A, nm	$M \times 10^{-3}$	Q
Poly-γ-benzyl-l-glutamate	240	33–330	0.88–0.38
Polybutyl isocyanate	100	15–150	0.76–0.19
Poly-hlorohexyl isocyanate	48	24–240	0.57–0.10
Poly-para-benzamide	110	5.5–55	0.45–0.18
Poly-para-phenyleneterephthalamide	65	5.5–55	0.46–0.13
Poly-para-amide-hydrazide	45	5.5–55	0.39–0.09
Poly-meta-phenylenisophthalamide	5	5.9–59	0.1–0.01
DNA	90	58.3–583	0.72–0.17
Double-chain polyphenylensiloxan	20	31–310	0.34–0.04
Cellulose nitrate	23	17.3–173	0.37–0.05
Ethyl cellulose	20	14–140	0.33–0.05
Cellulose phenylcarbanilate	16	23–230	0.28–0.04
Polystyrene	2	12–120	0.04–0.004
Polyethylene	2	3.3–33	0.04–0.004
Polymethylmethacrylate	2	12–120	0.04–0.004

Table 1 shows that with an increase in the flexibility of macromolecules (reduction of Khun's segment size) and in the polymer molecular mass, the value of the intermolecular orientation order decreases. It is important to emphasize that the transitions of the rigid-chain polymers into the LC state usually occur in solutions rather than in melts.

1.3 Effect of magnetic field on liquid crystalline systems

Low-molecular-mass liquid crystals exhibit a good orientation in the magnetic and electric fields. In general, this is associated with the anisotropy of diamagnetic susceptibility $\Delta\chi$ or dielectric permittivity $\Delta\varepsilon$ parallel and perpendicular to the long axis of molecules. The molecules of LC substances are ordered so that the direction with the higher values of χ (or ε) is parallel to a vector of field intensity [1]. However, the behavior of polymer LC systems in the force fields has not been studied enough.

The theory of the interaction of diamagnetic macromolecules with a magnetic field is in the development stage [9, 10, 11, 12, 13]. If such a macromolecule is placed in a magnetic field, then the force acting on it will cause it to rotate. This is due to the magnetic anisotropy of the molecule, depending on the magnetic anisotropy of chemical bonds. In these systems, an orientation of chains proceeds cooperatively.

The magnetic field leads to the orientation of macromolecular domains in a certain predominant direction that depends on the sign of diamagnetic susceptibility anisotropy ΔχM for this polymer. Domains are taken to mean the anisotropic associates of macromolecules. The diamagnetic moment appearing at the domain can be written as [9, 10]:

$$\mu = \Delta\chi^{M} B^{2}\ VSin2\xi / 2\mu_{0} \qquad (1)$$

where V is domain volume, μ0 is a magnetic constant of vacuum, B is a vector of magnetic induction, ξ is an angle between the direction B and the domain axis.

Interaction of external magnetic field with the domain having the magnetic moment μ increases energy of magnetic field by value of Emag. Orientation is observed when Emag exceeds the amount of thermal energy. In this case, the following conditions must be met [9, 11]: first, the particle must be anisodiametric;

secondly, the volume of the particle must be greater than the critical value.

$$Vcrit = 2kT\mu_{0} / B^{2} I\Delta\chi I, \qquad (2)$$

where μ0 is the magnetic constant of vacuum, B is the vector of magnetic induction, Δχ is the anisotropy of diamagnetic susceptibility, k is the Boltzmann constant, and T is the absolute temperature; third, the medium should be low viscosity.

The influence of the magnetic field on the liquid crystals was studied by Mayer [14] and de Gennes [15]. They examined the behavior of a cholesteric liquid crystal under magnetic field and found that at the critical intensity Hc of magnetic field, the complete transition to the nematic LC structure is observed. And

$$H_c = \frac{\pi^2}{2}\left(\frac{K_{22}}{\Delta\chi_m}\right)^{1/2}\frac{1}{p_0} \qquad (3)$$

where p0 is the pitch of the cholesteric helix before exposure of the magnetic field, Δχm is the anisotropy of magnetic susceptibility of the liquid crystal, K22 is the module of the resilience of cholesteric mesophase. This equation takes into account the anisotropy Δχ of one molecule. The theory predicts the slow increase in the pitch of the cholesteric helix at the small intensity of the magnetic field and the exponential increase in the pitch of the cholesteric helix near the critical intensity of the magnetic field. The theory has been checked for the lyotropic PBG liquid crystals in a number of solvents [16, 17]. According to Miller et al. [18], the molecules of liquid crystals are oriented parallel to the magnetic field lines. This orientation is related to the anisotropy of the magnetic susceptibility of molecules, which in turn is associated with the anisotropy of their structure. The effect of a magnetic field on the properties of polymer systems is discussed in [19, 20, 21, 22, 23, 24, 25, 26, 27, 28].

Since 2006, researchers of the Chair of Macromolecular Compounds, Ural State University investigate the effect of magnetic field on the phase transitions, structure and rheological properties of liquid crystalline solutions of cellulose ethers [12, 13, 29, 30, 31, 32, 33, 34, 35, 36].

2. Phase diagrams of solutions of rigid-chain polymers

A qualitative approach to examination of the solutions of rigid-chain polymers with rigid rod-shaped molecules reveals that an independent position (a random orientation) of each molecule is possible only in relatively diluted solutions. As the number of macromolecules in the given volume of solution increases, the probability of the random orientation of the rigid rod-shaped molecules decreases, and as soon as the polymer concentration reaches a certain critical value, the further increase in the number of molecules in the volume proves to be impossible without the relative ordering of part of them. Thus, it should be followed by separation in two phases, one of which exhibits ordered macromolecules and the other that ensures that molecules are randomly oriented relative to each other. As the concentration of polymer continues to increase, the share of the ordered phase grows, and eventually the system becomes one-phased again while all macromolecules are relatively ordered [1, 14, 29, 30, 31, 32, 33, 34, 35, 36, 37].

The first molecular theory of nematic ordering was offered by Onsager in 1949 [38] for a solution of long rigid cylindrical rod-shaped molecules of length L and diameter ($L \gg d$). Such a system is a simulation of the solution of extremely rigid-chain macromolecules with flexibility too small to be noticed along the L length. Onsager examined a case of an athermic solution where only repulsive forces exist between rod-shaped molecules due to their relative impermeability and the liquid crystalline ordering is caused by the steric factors. He revealed that:

1. Orientation order in solutions of long rigid rod-shaped molecules is a phase transition of the first order occurring at a low concentration of rod-shaped molecules in solution ($\varphi 2 \sim d/L \ll 1$);

2. At $\varphi 2 < \varphi 2i$, the solution is isotropic; at $\varphi 2 > \varphi 2a$, it is anisotropic; and at $\varphi 2i < \varphi 2 < \varphi 2a$, the solution separates into isotropic and anisotropic phases, where

$$\varphi_{2i} = 3.34(d/L); \varphi_{2a} = 4.49(d/L) \quad (4)$$

3. Order parameter $Q=\langle 3\cos 2\theta - 1\rangle/2$ in the point of LC phase appearance (i.e. at $\varphi 2 = \varphi 2a$) is equal to $Q = 0.84$.

Another approach to solving the problem of LC ordering in the solution of rigid rod-shaped molecules was developed by Flory in 1956 based on the lattice theory of solutions [39]. He introduced the following equation to describe free energy change:

$$\triangle G_{sm}/(RT) = n_1 \ln \varphi_1 + n_2 \ln \varphi_2 - (n_1 + yn_2) \ln [1 - \varphi_2(1 - y/x)] \\ - n_2 [\ln (xy_2) - y + 1] + \chi_1 x n_2 \varphi_1, \quad (5)$$

where φ1 and φ2 are the volume fractions of solvent and polymer, respectively; n1 and n2 are the quantity of molecules of solvent and polymer, respectively; χ1 is the Flory-Huggins interaction parameter; y is the parameter of macromolecule disorientation; x is the degree of molecule asymmetry (x = L/d, L is the length of a molecule, d is the diameter of a molecule).

As a result of the minimization of free energy of the system ΔGsm, it was revealed [39] that critical volume fraction of polymer φ_2^*, at which LC phase appears at x > 10 accurate within 2%, is equal to:

$$\varphi_2^* \cong \frac{8}{x}\left(1 - \frac{2}{x}\right) \quad (6)$$

Based on Flory's theory, a phase diagram of rigid-chain polymer solution was constructed.

The limits of the separation region in the case of athermic solution at x ≥ 200 were described by Flory as follows:

$$\varphi_i = 8/x, \ \varphi_a = 12.5/x. \quad (7)$$

Flory's lattice theory, despite its artificial character, can be successfully applied to the solving of a number of certain problems when other approaches require too complicated computations. The examples of phase diagrams based on experimental results are given below.

The phase diagram of PBG-DMF systems, experimentally found by Miller et al. [40] by methods of polarizing microscopy, viscometry, and nuclear magnetic resonance (NMR), exhibits the rather close agreement with Flory's theoretical diagram as shown in Figure 5. The experimental diagram represents a dashed region having been defined with low accuracy (Figure 6) [40]. According to Flory's calculations, this region testifies to the coexistence of two anisotropic phases.

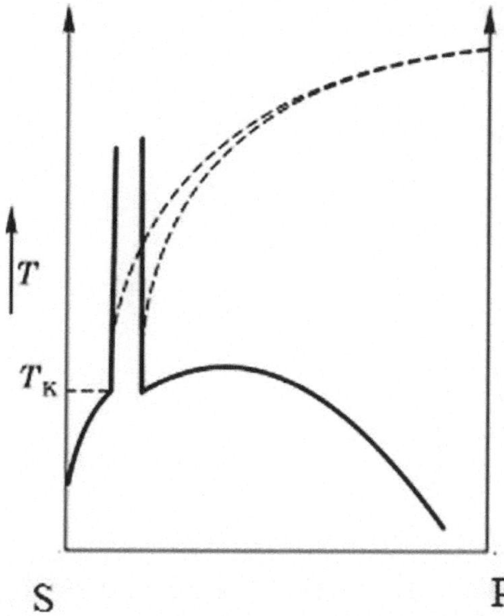

Figure 5. Phase diagram for a system of rigid-chain polymer-solvent based on Flory's theory (dashed line shows a deviation from the theoretical behavior corresponding to the temperature effect on the phase composition).

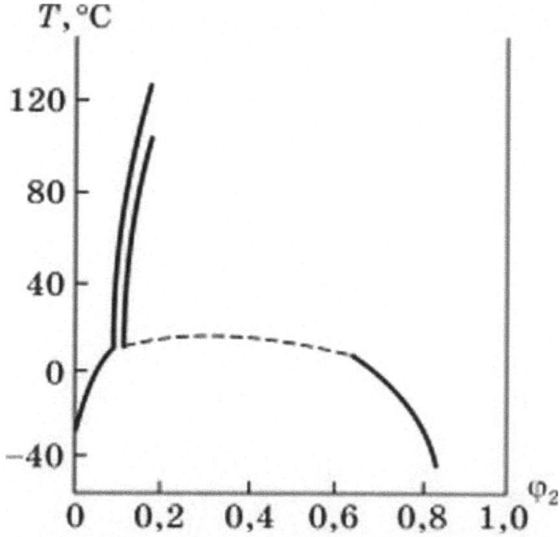

Figure 6. Phase diagram of poly-y-benzyl-l-glutamate (PBG)-dimethylformamide (DMF) system [40].

In the diagram of poly-carbobenzoxylysine-DMF system (Figure 7) [40], the narrow region of the coexistence of isotropic and anisotropic phases is observed, which becomes broader only at very low temperature values. Though the rather high freezing point of DMF prevents obtaining experimental values in the region where the compositions of coexisting phases approach pure solvent and pure polymer correspondingly, the tendency toward transition to this area is quite obvious.

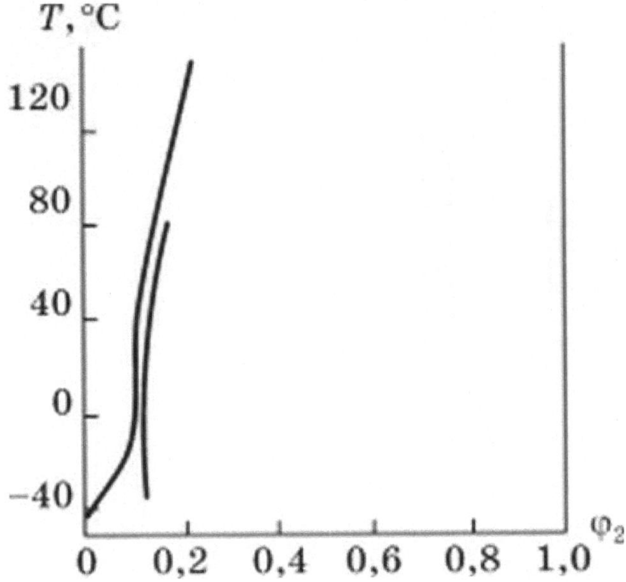

Figure 7. Phase diagram of polycarbobenzoxylysine-DMF system [40].

The diagram in Figure 8 was obtained by Iovleba et al. [41]. One can see that it represents the part of the theoretical diagram associated with the narrow region of the coexistence of isotropic and anisotropic phases.

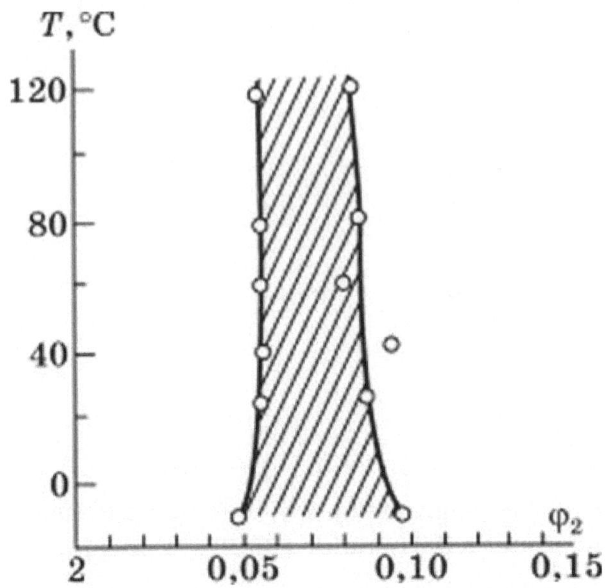

Figure 8. Phase diagram of poly-p-benzamid (PBA)-dimethylacetamide (DMAc) systems (+3% mass. LiCl) [41].

The phase diagrams were constructed for the following systems: PBA-DMAc [42], PBG-benzyl alcohol [43], PPTA (poly-p-phenylene terephtalamid)-H2SO4-water [44], PBG-m-cresol [45], copolymer of p-phenylenediamine with terephthalic acid-H2SO4, and copolymer of p-phenylene-diamine with 4,4′-(diphenyl dicarbonic) acid-H2SO4 [46], PBG-dichlorine acetic acid [47].

While examining polymer LC systems, one cannot but refer to the studies dedicated to exploring the LC states of cellulose and its derivatives [48]. Cellulose is one of the most widespread natural polymers used intensively in various industries. The liquid crystalline state in solutions and melts of several cellulose derivatives was discovered in the 1980s. A number of studies of molecular structure of cellulose and its derivatives revealed that they have a helical conformation in the ordered regions being stabilized by intermolecular hydrogen bonds. If during the dilution of the aforesaid polymers the intermolecular H-bonds remain, the molecules continue to be rigid-chained and therefore are able to produce the ordered structures and form a mesophase. If during the dilution the intermolecular H-bonds are disrupted, the macromolecules become flexible and as a result cease to be ordered. Thus, to maintain the LC solution of cellulose and its derivatives, so-called chiral forming solvents are used to disrupt only the intermolecular H-bonds (DMF, DMAC, 1,4-dioxane, chlorinated hydrocarbons) [48].

Figures 9 and 10 show the phase diagrams of solutions of cellulose derivatives.

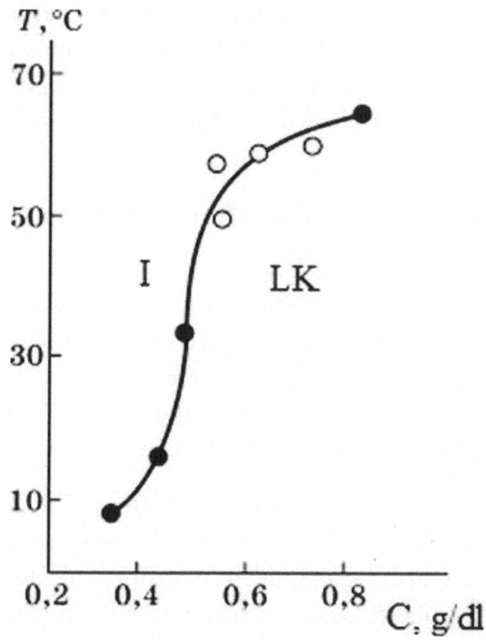

Figure 9. Segment of the phase diagram of the cellulose tricarbonylate-methyl ethyl ketone system [49].

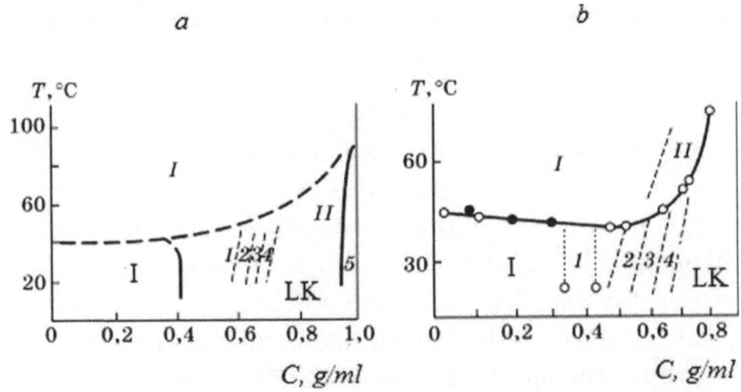

Figure 10. Phase diagram of the hydroxypropyl cellulose (HPC)-water system [50] (a) and [51] (b): (1) turbid solution in case (a) and two-phase system in case (b); (2) red; (3) green; (4) violet; (5) transparent film. (I) Precipitation; (II) gel.

Systematic studies of liquid crystalline phase transitions in solutions of cellulose esters were carried out at the polymer chair of Ural Federal University [12, 13, 29, 30, 31, 32, 33, 34, 35, 36]. The objects of this study were hydroxypropyl cellulose (HPC) of the Klucel-JF trademark (Hercules, United States); cyanoethyl cellulose (CEC), cyanoethylnitrocellulose (CENC), cyanoethyl, hydroxyethyl cellulose (CEHEC), and methyl cellulose (MC) (Institute of Macromolecular Compounds, Russian Academy of Sciences, Russia); and hydroxyethyl cellulose (HEC) and ethyl cellulose (EC) (Hercules-Aqualon and ACROS–USA). Their characteristics are presented in Table 2.

2.1 Effect of polymer molecular mass on phase LC transitions in cellulose ether-solvent systems

According to [39], the critical volume fraction of a polymer φ^* at which the LC phase is formed is related to the asymmetry of a macromolecule x via the equation $\varphi^* = (1-2/x)8/x$, where x is the degree of asymmetry of the macromolecule (the length-to-diameter ratio of the molecule). The greater the molecular mass of the macromolecule, the higher the x and the lower the φ^*.

Figure 11 shows the boundary curves for the HPC-1-DMAA, HPC-3-DMAA, HPC-1-ethanol, HPC-2-ethanol, HPC-3-ethanol, HEC-1-water, and HEC-3-water systems. These curves separate transparent isotropic solutions (I) from opalescent anisotropic solutions (II) ($\omega 2$ is the mass fraction of the polymer). It is shown that a rise in polymer molecular mass leads to a shift of the boundary curve to the region of lower concentrations. It should be noted that the Flory' theory does not consider influence of the chemical structure of polymer and solvent molecules on phase LC transitions.

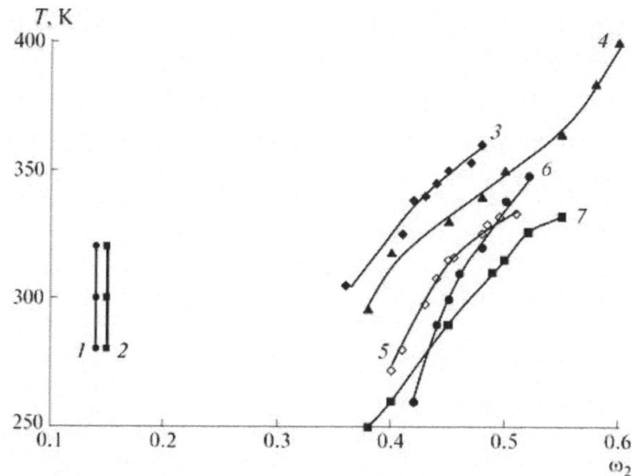

Figure 11. Boundary curves of (1) HEC-4 (Mη = 4.5 × 105)-water, (2) HEC-1 (Mw = 6.2 × 104)-water, (3) HPC-4 (Mw = 1.15 × 106)-ethanol, (4) HPC-2 (Mw = 1.5 × 105)-ethanol, (5) HPC-4 (Mw = 1.15 × 106)-DMAA, (6) HPC-1 (Mw = 0.9 × 105)-ethanol, and (7) HPC-1 (Mw = 0.9 × 105)-DMAA systems [52].

2.2 Effect of component nature on liquid-crystalline transitions in solutions of cellulose ethers

2.2.1 Effect of the chemical structure of polymer molecules on phase LC transitions

Figure 12 illustrates the data on phase transitions for the HPC-1-DMF and EC-1-DMF systems [52]. It is seen that the boundary curve of the EC-1-DMF system is in the region of lower concentrations than the boundary curve of the HPC-1-DMF system. Replacing the branched hydroxylpropyl radical with the ethyl radical enhances interaction between the units of neighboring macromolecules of the EC polymer, thereby facilitating formation of the LC phase. A lower polarity of EC-1 molecules worsens interaction with the polar DMF solvent.

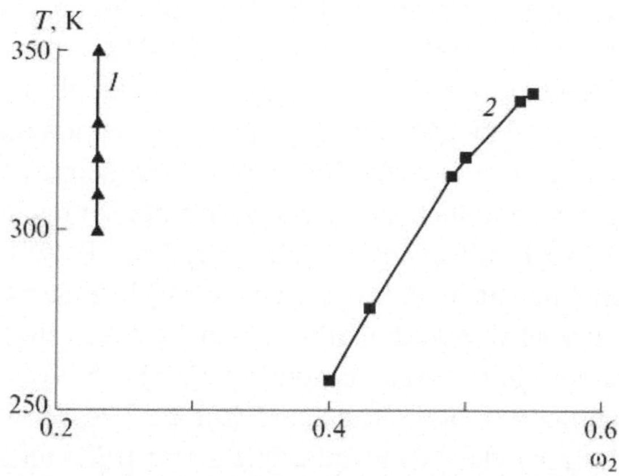

Figure 12. Boundary curves of (1) EC-1(Mη = 2.6 × 104)-DMF and (2) HPC-1(Mw = 0.9 × 105)-DMF systems.

Figure 13 shows the boundary curves of the EC-1-ethanol, EC-1-DMF, HPC-ethanol, and HPC-2-DMSO systems. It is shown, that, as for DMF solutions in Figure 12, in the case of ethanol solutions, the replacement of the hydroxypropyl radical with the ethyl radical in cellulose ether units causes a decrease in the concentration of formation of the LC phase.

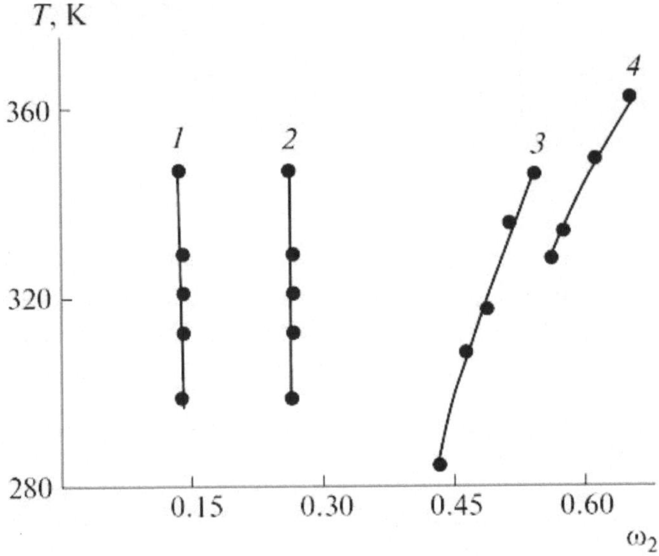

Figure 13. Boundary curves of (1) EC-1(Mη = 2.6 × 104)-ethanol, (2) EC-1(Mη = 2.6 × 104)-DMF, (3) HPC-2 (Mw = 1.5 × 105)-ethanol, and (4) HPC-2 (Mw = 1.5 × 105)-DMSO systems.

A similar dependence is shown for the EC-1-DMF and HPC-2-DMSO systems. The boundary curves for the HPC-ethanol, HPC-DMAA, and HPC-DMSO systems change with temperature (Figures 11–13). At high temperatures the thermal motion of the molecules can disturb the liquid-crystalline order. Consequently, a high concentration of the polymer is required to retain the liquid-crystalline order. The boundary curves of the HEC-water (Figure 11), EC-DMF, and EC-ethanol systems (Figure 13) do not depend on temperature and are located in the region of lower concentrations. It is caused by a strong interchain interaction and a higher packing density of these macromolecules. The linear ethyl and hydroxyethyl radicals in the units of neighboring macromolecules can produce a denser packing with each other than the branched propyl radicals of HPC [53, 54]. The IR study of cellulose ether films showed [55] the formation of intra- and intermolecular hydrogen bonds with different strengths. It was shown that HEC is a more associated polymer than CEC and CEHEC. CEC is a less associated compound. CEHEC occupies the intermediate position. Methyl cellulose is also associated but to a lesser extent. As a result, large supramolecular particles are formed in EC and HEC solutions, which do not decompose under heating (Table 3).

Table 2. Characteristics of research objects.

Cellulose ether	Molecular mass	Degree of substitution, α
HPC-1	$M_w = 0.9 \times 10^5$	3.0
HPC-2	$M_w = 1.5 \times 10^5$	3.3
HPC-3	$M_\eta = 4.5 \times 10^5$	3.0
HPC-4	$M_w = 1.15 \times 10^6$	3.0
CEC	$M_w = 0.90 \times 10^5$	2.6
HEC-1	$M_w = 6.2 \times 10^4$	2.5
HEC-2	$M_w = 8.6 \times 10^4$	2.5
HEC-3	$M_\eta = 1.0 \times 10^5$	2.5
HEC-4	$M_\eta = 4.5 \times 10^5$	2.5
EC-1	$M_\eta = 2.6 \times 10^4$	2.6
EC-2	$M_w = 1.6 \times 10^5$	1.5
CENC	[η] = 2.79 dL/g (DMAA, T = 298 K)	0.19
MC	$M_\eta = 1.8 \times 10^5$	2.0
CEHEC	$M_\eta = 1 \times 10^5$	2.0

Table 3. Diameters dw of light-scattering particles and mean-square distances between chain ends (h2)1/2 of cellulose ether macromolecules, T = 298 K.

System ($\omega_2 = 0.05$)	d_w, nm	$(h^2)^{1/2}$, nm
CEC-DMAA	110	60
CEC-DMF	300	60
HPC-1-ethanol	320	64
HPC-1-water	180	64
HEC-1-water	1780	45
HEC-1-DMF	2480	45
HEC-2-DMF	1600	51
HEC-2-DMAA	1800	51
EC-2-DMAA	580	80

The value of dw = 2rw, where rw is the weight-average radius of particles scattering light, as determined by the turbidity spectrum method (this is described below). A comparison of the values of dw and the sizes of macromolecules (h2)1/2 (Table 3) shows that, at a mass fraction of the polymer of ω2 = 0.05, light-scattering particles are supramolecular. The light-scattering particles of HPC and CEC consist of several macromolecules. The molecular-disperse solutions are not formed in the systems based on HEC and EC. Coarse light-scattering particles are probably the residues of the initial structure of the polymer stabilized by a great amount of strong hydrogen bonds formed between neighboring hydroxyl groups and a denser packing of linear ethyl and hydroxyethyl radicals in the units of neighboring macromolecules. Therefore, the position of the boundary curves for the EC-DMF and EC-ethanol systems does not depend on temperature.

Figure 14 shows the boundary curves of the CENC-DMAA, CENC-DMF, CEC-DMF, and CEC-DMAA systems. It is shown that, in the CENC solutions, the LC order appears at a lower concentration of the polymer. The introduction of a nitro group into CEC chain units increases the rigidity of chains; so the degree of asymmetry of macromolecules grows and the LC phase is formed at lower concentrations of the polymer.

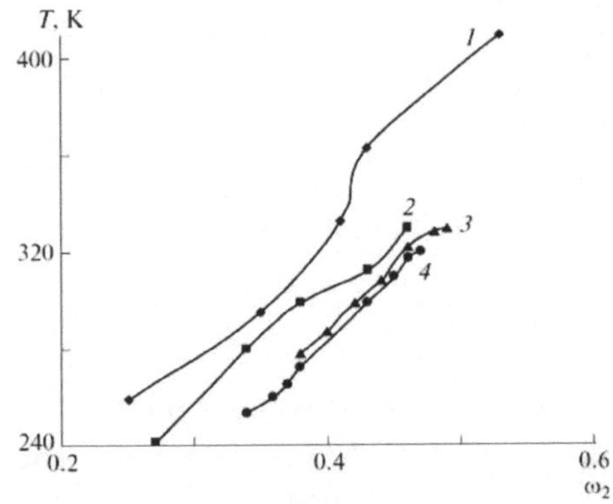

Figure 14. Boundary curves of (1) CENC-DMAA, (2) CENC-DMF, (3) CEC (Mw = 0.9 × 105)-DMF, and (4) CEC (Mw = 0.9 × 105)-DMAA systems [52].

The boundary curves for the aqueous solutions of HPC-1 and HEC-1 are given in Figure 15.

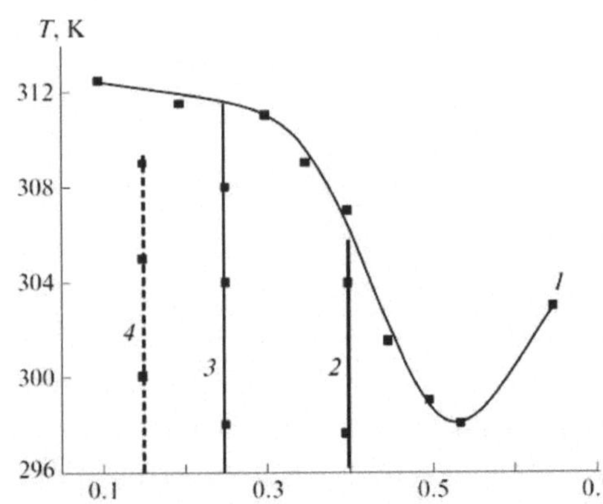

Figure 15. Boundary curves of (1–3) HPC-1(Mw = 0.9 × 105)-water and (4) HEC-1 (Mw = 6.2 × 104)-water systems. Lines 3 and 4 are the boundaries of the isotropic-anisotropic solution transition, and lines 2 and 3 restrict the region of coexistence of isotropic and anisotropic phases [52].

Obviously that the replacement of the hydroxypropyl radical with the hydroxyethyl one causes a substantial change in the phase diagram: first, the LC phase in HEC solutions is formed at lower concentrations. This is caused by the linear character and a smaller size of the hydroxyethyl radical compared with the hydroxypropyl one. So, interchain interaction is enhanced. Secondly, the LCST is absent in the HEC solutions. HPC contains 10–15% of the crystalline phase and 90–85% of the vitrified cholesteric LC phase [56, 57]. The degree of crystallinity of HEC is almost zero. Structural changes during dissolution of HPC in water may be caused by the melting of crystalline domains and relaxation of the metastable glass-like structure. For HEC, only relaxation of the metastable glass-like structure can be observed. The interactions of HPC with water are determined both by the salvation of macromolecules by water owing to the formation of hydrogen bonds [58] and the "hydrophobic hydration" of water itself [59]. The latter process includes the compaction of water structure during penetration of nonpolar molecules or their fragments into voids in its structure. Methyl and methylene groups of HPC are nonpolar fragments. Therefore, intermolecular distances in water decrease, the interaction between its molecules becomes stronger, the exothermic effect of dissolution grows, and the entropy of mixing decreases [60]. The interactions of both types contribute to the negative enthalpy and entropy of mixing of HPC with water. Phase separation of the HPC-water system during heating may be caused by the melting of the densified water structure around hydrophobic fragments of the polymer [61]. The interactions of HEC with water can also be determined both by the solvation of macromolecules by water and by the "hydrophobic hydration" of water itself. However, since there are no methyl groups in HEC molecules to a lesser extent the LCST is absent for this system.

Figure 16 shows the boundary curves of the HEC-2-DMAA, EC-DMAA, and HPC-1-DMAA systems.

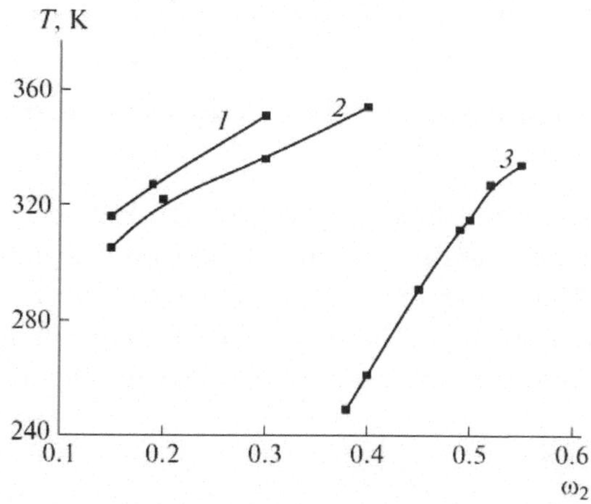

Figure 16. Boundary curves of (1) HEC-2 (Mw = 8.6 × 104)-DMAA, (2) EC-2 (Mw = 1.6 × 105)-DMAA, and (3) HPC-1 (Mw = 0.9 × 105)-DMAA systems [52].

The solutions of HEC, EC, and HPC in DMAA follow the same shift in the boundary curves as water and ethanol solutions because of the reasons described above. The replacement of the hydroxyethyl radical in HEC with the ethyl radical in EC causes the weakening of interaction owing to a decrease in the possible formation of hydrogen bonds and, as a result, owing to a rise in concentration of LC phase formation. Note that the boundary curves for the solutions of HEC and EC are in the region of lower concentrations than the solutions of HPC. This is connected with the presence of bulky branched hydroxypropyl radicals in molecules. The boundary curves for the HEC-1-DMF, CENC-DMF, CE-DMF, and HPC-1-DMF systems are presented in Figure 17. For the solutions of these polymers in DMF, the same regularities as for the solutions in DMAA are observed.

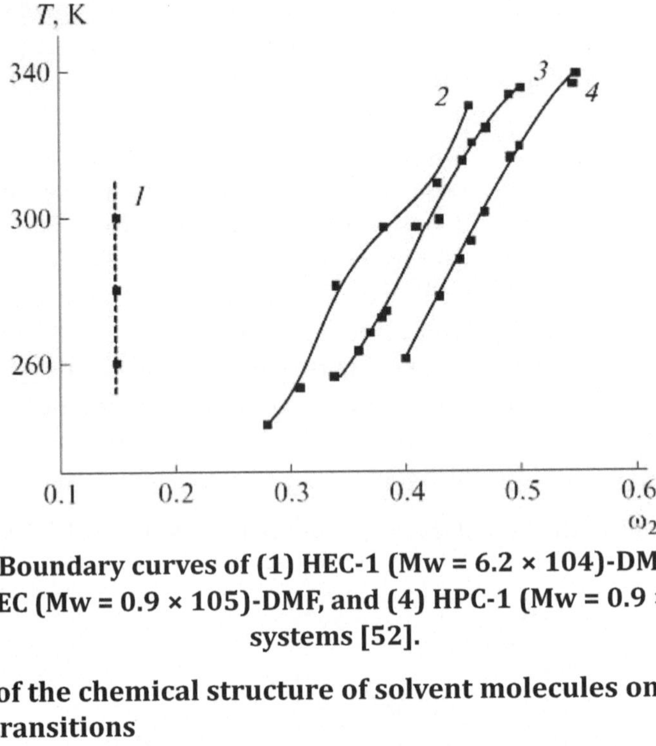

Figure 17. Boundary curves of (1) HEC-1 (Mw = 6.2 × 104)-DMF, (2) CENC-DMF, (3) CEC (Mw = 0.9 × 105)-DMF, and (4) HPC-1 (Mw = 0.9 × 105)-DMF systems [52].

2.2.2 Effect of the chemical structure of solvent molecules on phase liquid crystalline transitions

Figures 18–20 show the boundary curves for HPC and CEC solutions. The position of the boundary curves on the composition significantly depends on the solvent nature. The better the solvent, to a lesser degree it deteriorates the initial structure of the polymer. Therefore, the LC phase in solution will form at a higher polymer concentration. A more polar solvent should be better for polar cellulose ethers.

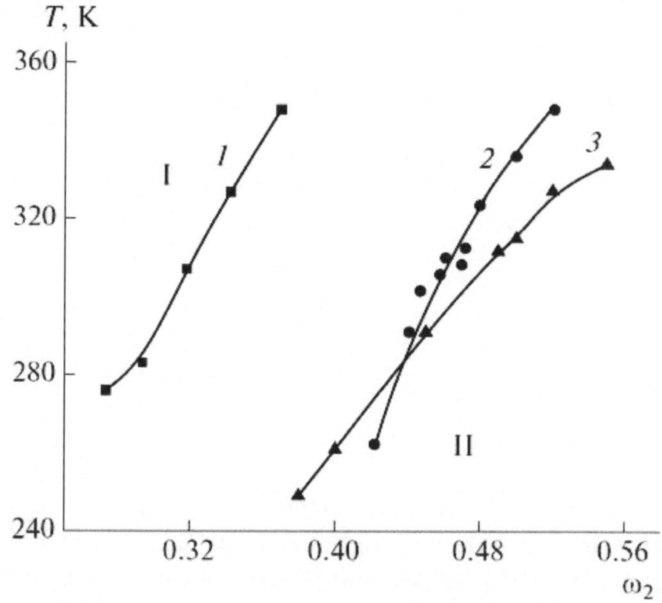

Figure 18. Boundary curves of (1) HPC-1 (Mw = 0.9 × 105)-acetic acid, (2) HPC-1 (Mw = 0.9 × 105)-ethanol, and (3) HPC-1 (Mw = 0.9 × 105)-DMAA systems [52].

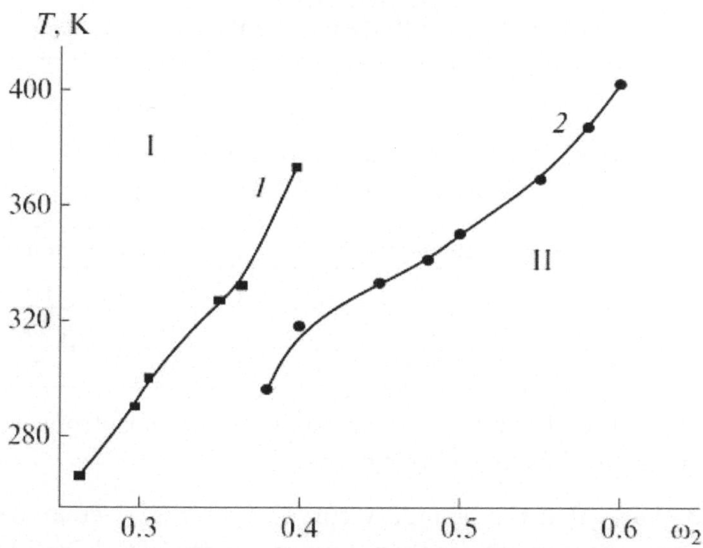

Figure 19. Boundary curves of (1) HPC-2 (Mw = 1.5 × 105)-acetic acid and (2) HPC-2 (Mw = 1.5 × 105)-ethanol systems [52].

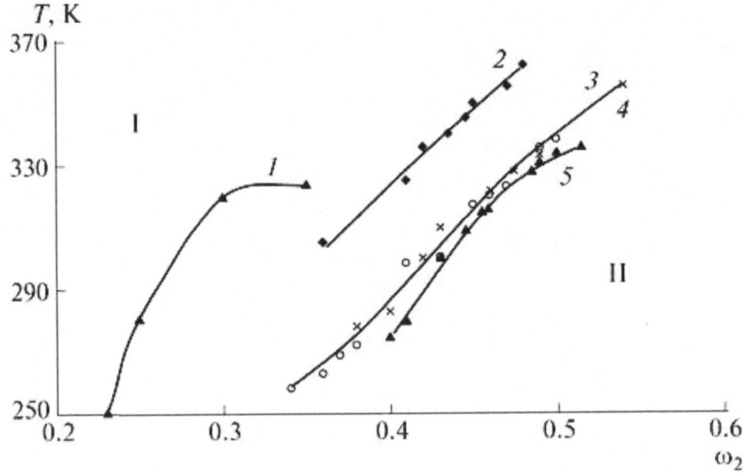

Figure 20. Boundary curves of (1) CEC (Mw = 0.9 × 105)-TFAA/MEC, (2) HPC-4 (Mw = 1.15 × 106)-ethanol, (3, light circles) CEC (Mw = 0.9 × 105)-DMF, (4) CEC (Mw = 0.9 × 105)-DMAA, and (5) HPC-4 (Mw = 1.15 × 106)-DMAA systems (TFAA/MEC is trifluoroacetic acid/methylene chloride, 1/1 volume fractions) [52].

Table 4 compares the second virial coefficients and dipole moments of solvent molecules. An increase in the polarity of solvent molecules entails a rise in second virial coefficients, suggesting the improvement of thermodynamic interaction of the components.

Table 4. Second virial coefficients and dipole moments of solvent molecules for cellulose acetate solutions [62].

Solvent	A_2×10^4cm^{3}mol/g^2	Dipole moment, D
DMAA	7.5	3.86
Water	3.1	1.68
Formamide	8.4	3.7

Here and in Figures 19 and 20, I refers to the region of isotropic solutions, and II refers to the region of anisotropic solutions.

Figure 13 shows that the boundary curve of the HPC-2-ethanol system is in the region of lower concentrations than the curve of the HPC-2-DMSO system. This may be caused by different polarities of solvent molecules. The dipole moment of ethanol molecules μ = 1.69 D [63] is lower than the dipole moment of DMSO μ = 3.96 D [63]. A more polar solvent DMSO destroys the initial structure of the polar polymer to a greater extent; therefore the sizes of supramolecular particles decrease and a higher concentration of polymer is needed for formation of the LC order in solutions. A similar behavior is observed for the EC solutions. The dipole moment of DMF molecules μ = 3.81 D [63] is higher than the dipole moment of ethanol molecules μ = 1.69 D, and the LC state forms in the EC -DMF system at a higher polymer concentration. Table 5 compares the thermodynamic parameters of interaction for the cellulose ether-solvent systems and the physical properties of the solvents [60, 63].

Table 5. Dipole moments and minimum Gibbs energies of mixing of cellulose ethers with low-molecular-mass liquids.

Polymer	Solvent	μ, D [63]	$-\Delta g^m$, J/mol solution [60]
MC	Chloroform	1.06	1400
	H_2O	1.83	1350
	Ethanol	1.69	1200
CEC	Chloroform	1.06	1400
	H_2O	1.83	1250
	Acetone	2.85	1400
	Ethanol	1.68	900
CEHEC	Chloroform	1.06	1400
	TFAA	2.28	4250
	H_2O	1.83	1150
	Dioxane	0.45	1250
	Acetone	2.85	1700
HEC	Chloroform	1.06	1300
	H_2O	1.83	1700
	Ethanol	1.68	1050
	Dioxane	0.45	300

The dissolving capacity of water is due to the possibility to form hydrogen bonds between its molecules and molecules of polymers. For HEC containing the maximal amount of hydroxyl groups, water is the best solvent.

The good interaction with chloroform and water is characteristic of MC, the least associated polymer with a small nonpolar substituent and unsubstituted hydroxyl groups. CEC and CEHEC containing polar nitrile groups with donor properties [64] have a high affinity for chloroform and a low affinity for water despite the presence of unsubstituted hydroxyl groups. Owing to a high packing density and, as a result, strong interchain interaction, CEC does not dissolve but only limitedly swells in water. This may suggest an insignificant role of hydrogen bonds in dissolving cyanoethylated ethers. This assumption is confirmed by a low affinity of CEC for ethanol. Dioxane being the electron donor for polymer molecules [64] capable of H bonding has a low thermodynamic affinity for HEC and CEHEC.

With some exceptions, the higher the dipole moment of solvent molecules, the more negative the Gibbs energy of mixing, i.e., the better the solubility of the cellulose ethers. Thus, in a first approximation, the dipole moment can be used as an estimate of solvent quality for cellulose ethers.

3. Self-assembly of nanodimensional molecules of cellulose ethers

An important specific task to be solved within the priority direction of science and technology—industry of nanosystems and materials—as well as in critical technology—nanotechnologies and nanomaterials—is the study of the process of self-assembly of supramolecular systems. Such systems include the solutions of rigid chain polymers, the molecules of which exhibit an ability for self-organization leading to the formation of liquid crystalline phases. Informative methods employed for the research of solution structures are the optical methods: the method of Rayleigh light scattering, the method of dynamic light scattering, the method of measurement of optical density (turbidimetry). The chapter presents the results of the examination of solution structure using the turbidity spectrum method, which makes it possible to determine the size of supramolecular particles in solutions in a wide composition range. The method was introduced by Heller et al. [65, 66, 67] and further developed by Klenin et al. [68]. According to this method, the typical dependence of logarithm of optical density A on the logarithm of wavelength required for computation of rw is described by straight line. The decrease of A with the increase of wavelength λ is observed, which is in agreement with Angstrom's Eq. A ⍰ λ−n. Wavelength exponent n depends on the size of light scattering particles and is connected with parameter α = 2πrw/λaverage and relative refractive index mrel = ndpolymer/ndsolvent. Value λaverage = λo/ndsolvent (λ0 is the wavelength of light in vacuum corresponding to the middle of the linear section of the plot lnA = f(lnλ), rw is the weighted average of particle radius). The slope of the lines lnA = f(lnλ) was used to identify n and then α for the given mrel and rw [68].

Figure 21 shows the concentration dependencies of optical density A (for λ = 490 nm) experimentally obtained at 298 K for the systems: HEC-2-DMF and HEC-1-DMF. The dashed line separates the region of isotropic (I) solution from the region of anisotropic (II) solution. The similar dependences were obtained for HPC-1-ethanol, HPC-1-water, HEC-1-water, HEC-3-water, CEC-DMAc, HEC-2-DMAc, and EC-DMAc systems. These diagrams show that as the concentration increases, the optical density increases also. This indicates the process of structure formation in the systems, which must manifest itself through the increase in the size of light scattering particles rw. As is evident from the comparison with phase diagrams, the increase of A is the most abrupt at the LC phase formation.

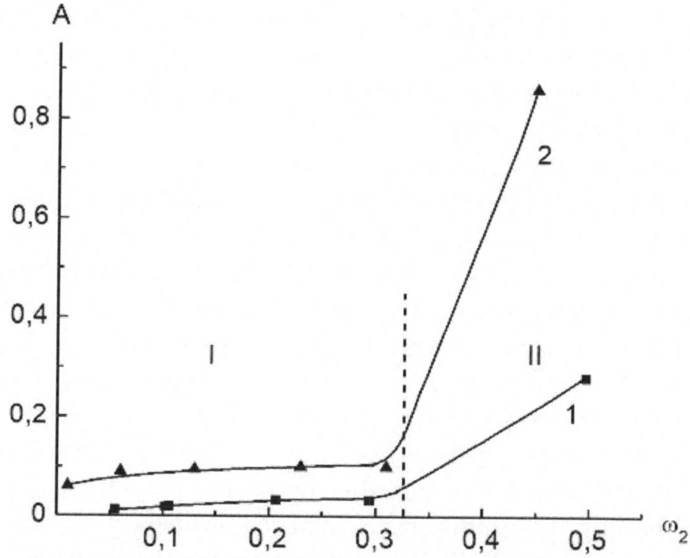

Figure 21. Concentration dependencies of the optical density for HEC-2-DMF (1) and HEC-1-DMF (2) systems: (I) isotropic solutions, (II) anisotropic solutions.

The concentration dependencies of the calculated average weighted radii of supramolecular particles at 298 K are given in Figure 22. The similar dependences were obtained for the HEC-1-water, HEC-3-water (2), HPC-water HEC-1-DMF, HEC-2-DMF, HEC-2-DMA, and EC-DMAc systems.

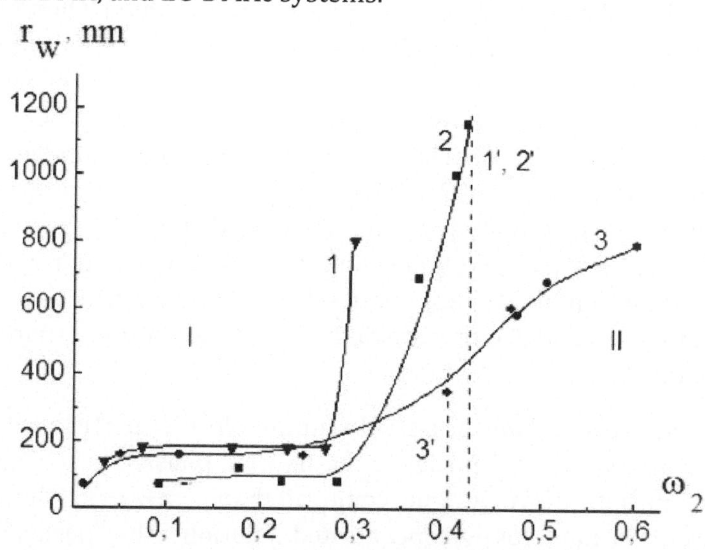

Figure 22. Concentration dependencies of the sizes of light scattering particles for the systems: CEC-DMF (1), CEC-DMAc (2), HPC-1-ethanol (3). (1′, 2′, 3′) the phase boundaries [12].

According to Table 3 and Figure 22 for the systems: CEC-DMAc, CEC-DMF, HPC-ethanol, HPC-1-water, HEC-3-water at polymer concentration of up to $\omega 2 \approx 0.05$, the diameters of light scattering particles do not exceed 320 nm, for the EC-DMAA system, they are of 580 nm and for the HEC-1-water, HEC-1-DMF, HEC-2-DMF, and HEC-2-DMAc systems, they vary from 1600 to 2500 nm. The calculated values (h2)1/2 show that the light scattering particles contain several macromolecules, except for the systems based on HEC where the molecular-dispersed solutions are unlikely to occur. According to studies [48], if the hydroxyl groups in cellulose macromolecules are completely substituted, their solubility at a molecular level is observed, while the partial substituted cellulose derivatives, as a rule, do not dissolve up to single macromolecules. The big light scattering particles are probably the residues of initial polymer structure stabilized by many strong hydrogen bonds between neighboring hydroxyl groups. After partial substitution, a certain portion of such bonds remains undisturbed by a solvent and large aggregations are found in the solutions. In [69] the size of light scattering particles was determined using the light scattering method. It was shown that in the solutions of samples of methyl hydroxypropyl cellulose and methyl hydroxyethyl cellulose, with M varying from 103 to 106 in water and cellulose acetate in acetone, the average number of aggregated molecules in the particles in diluted solutions grows from 10 to 103. In diluted water solutions of methylcellulose with a polymerization degree of 120, there are aggregations comprising 4000 macromolecules [70].

The sharpest increase of particle sizes is observed in the LC phase. It should be noted that for all studied systems, the $rw = f(\omega 2)$ curves exhibit a horizontal section in a certain concentration range indicating the constant size of supramolecular particles. The obtained data made it possible to suggest the following mechanism of self-assembly of cellulose ester macromolecules preceding LC transition. In diluted solutions there exist associates, including several macromolecules. Within the concentration range up to $\omega 2 \approx 0.05$, supramolecular particles of a stable-sized "packet" are formed. The term "packet" has been used in physics for a long time. For example, the wave packet is a certain summation of waves limited spatially and by time. Thus, in quantum mechanics, the description of a particle as a wave packet contributed to the acceptance of statistic interpretation of the square of absolute value of wave function.

Here the term "packet" means a stable supramolecular particle with essentially stable dimensions, which is a nuclei of the new LC phase. These supramolecular particles contain both polymers and conjugated solvents. As is shown in Figure 22, in the region of moderately concentrated solutions, the "packet" dimensions remain virtually unchanged within a certain concentration range, depending on the system. As the concentration of polymer in solutions further increases, the number of packets also increases while their size does not change. The transition of the system to a fully LC state is induced by the aggregation of packets with the formation of large particles thousands of nanometers in size. It is obvious that these large particles include a great number of macromolecules.

3.1 Effect of magnetic field on size of supramolecular particles in solutions of cellulose derivatives

To study the influence of magnetic field on the size of supramolecular particles, a cell with the polymer solution was placed in a gap between the electromagnet poles and kept at a field intensity of 9 kOe for 50 min. The optical density was measured within 5–10 min after the magnetic field was turned off.

It was shown that the EC-DMAc, HEC-1-DMF, and HEC-1-water systems exhibit an increase in optical density after magnetic-field treatment. The increase in the optical density of solutions is indicative of the occurrence of an additional association of macromolecules because they are oriented by their axes concerning lines of magnetic field.

It was revealed that the magnetic field leads to the increase in the radii of light-scattering particles in solutions. The concentration dependences of the sizes of supramolecular particles for the EC-DMAc, HEC-1-DMF, and HEC-1-water systems are presented in Figures 23–25.

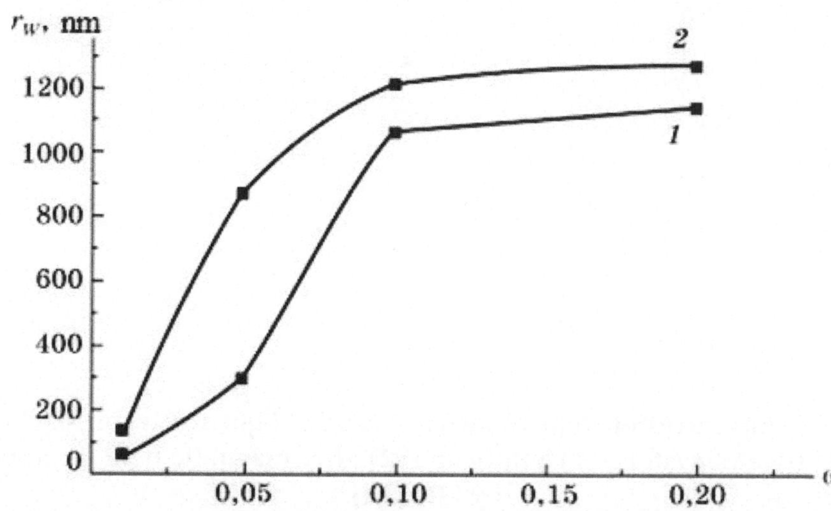

Figure 23. Concentration dependence of radii of light-scattering particles in the EC-DMAc system: (1) before and (2) after magnetic-field treatment. H = 9 kOe [12].

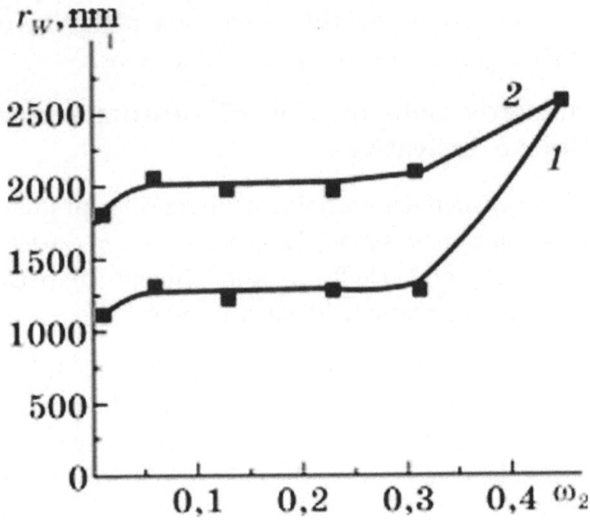

Figure 24. Concentration dependencies of radii of light-scattering particles in the HEC-1-DMF system: (1) before and (2) after magnetic-field treatment. H = 9 kOe [12].

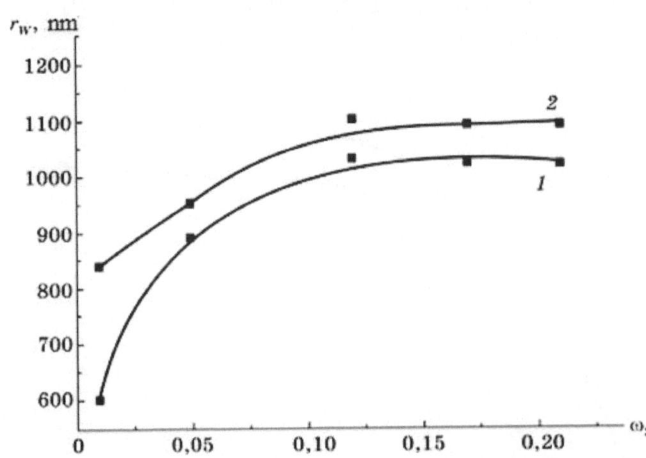

Figure 25. Concentration dependencies of radii of light-scattering particles in the HEC-1-water system: (1) before and (2) after magnetic-field treatment. H = 9 kOe [12].

4. Phase transitions of liquid crystalline systems in magnetic field

Experiments in the magnetic field were performed using a setup generating a constant magnetic field with a strength up 15 kOe [12]. A sealed ampoule with a polymer solution, which was transparent at elevated temperatures, was placed into a gap between magnet poles. The vector of the magnetic field strength was directed normal to the solution layer (~5-mm thick). The solution was cooled using a thermostating jacket, and the onset temperature of opalescence development was recorded. This temperature Tph was related to the appearance of the LC state.

The type of the phase transition was determined with the use of the polarization-photoelectric setup [12]. The sealed ampoule containing a transparent solution was placed into a gap between crossed polaroids (a polarizer and an analyzer). The ampoule was cooled using the thermostating jacket. A light beam from an helium-neon laser was transmitted through the polaroids in the direction normal to the ampoule containing the solution. When the solution was transparent (isotropic), the intensity of the transmitted light was zero. As the system became turbid upon cooling, the transmitted light intensity increased, as measured with a photoresistor. This fact suggested formation of an anisotropic phase, that is, the LC phase transition.

The phase diagrams were obtained under magnetic field for the systems: HPC-ethanol, HPC-DMAc, HPC-water, CEC-DMAc, CEC-DMF, PBG-DMF. The LC solutions of the cellulose esters exhibited the increase in the temperature of phase LC transition in the magnetic field and the preservation of the increased temperature for many hours after the magnetic field was switched off. This means that the cellulose ether-solvent systems are the "systems with memory." Figures 26–28 show the results of the experiments [12].

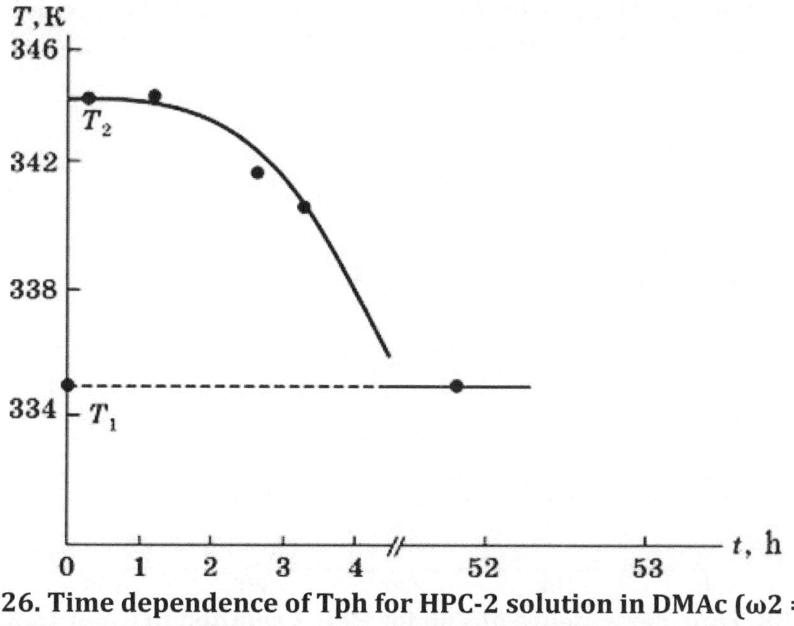

Figure 26. Time dependence of Tph for HPC-2 solution in DMAc (ω2 = 0.51) at T = 370 K after the influence of magnetic field with intensity of 7 kOe. T1 is the temperature of the LC transition without the field. T2 is the temperature of the LC transition under the field.

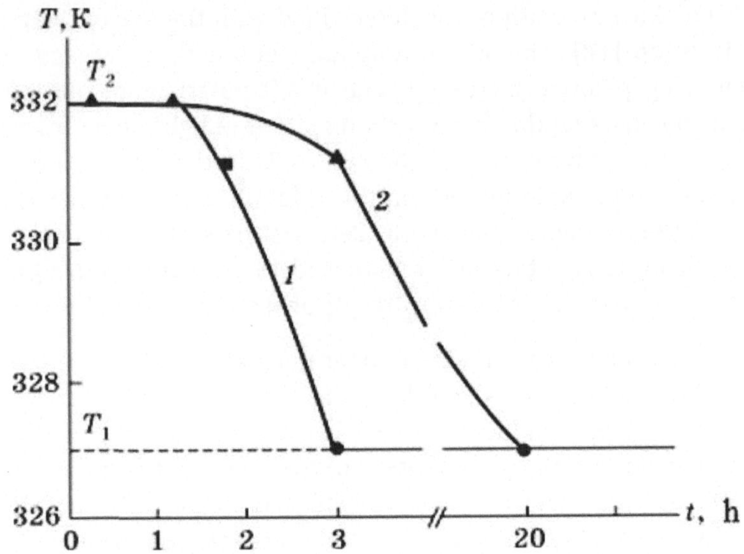

Figure 27. Time dependencies of Tph for CEC solution in DMAc (ω2 = 0.48) at T = 370 (1) and 298 K (2) after the influence of magnetic field with intensity of 7 kOe. T1 is Tph at H = 0 kOe, T2 is Tph at H = 7 kOe.

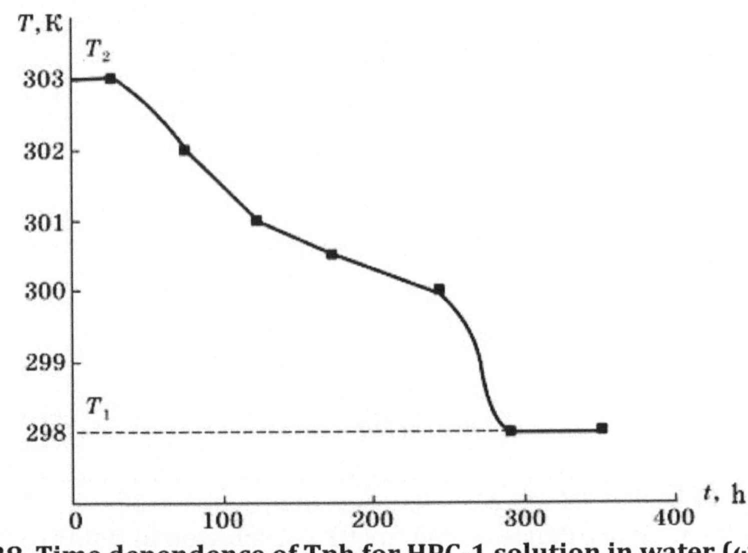

Figure 28. Time dependence of Tph for HPC-1 solution in water (ω2 = 0.54) at T = 298 K after the influence of magnetic field with intensity of 5 kOe. T1 is Tph at H = 0 kOe, T2 is Tph at H = 5 kOe.

It is observed that the magnetic field increases the temperature of LC phase transition. On the basis of the data from other studies, it seems to be induced by the change in the liquid crystalline type, in particular, by the LC cholesteric → LC nematic phase transition.

Figures 26–28 show that in the course of the time, the phase LC transition temperature increased by the magnetic field goes back down to the initial value (prior to the exposure to magnetic field). This means that the thermal motion destroys the field-induced orientation of macromolecules. The higher the temperature of thermostatic control of the solutions, the more intensive the thermal movement and the faster the recovery of the initial structure of solutions.

The boundary curves of the cellulose ether-solvent systems in the magnetic field are presented in Figures 29–32 [12] (I is the region of isotropic solutions, II is the region of anisotropic solutions).

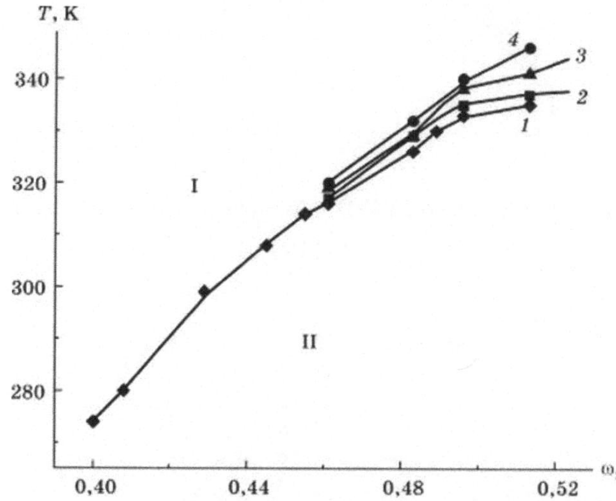

Figure 29. Boundary curves of HPC-1-ethanol. H = 0 (1), 3 (2), 5 (3), and 9 (4) kOe.

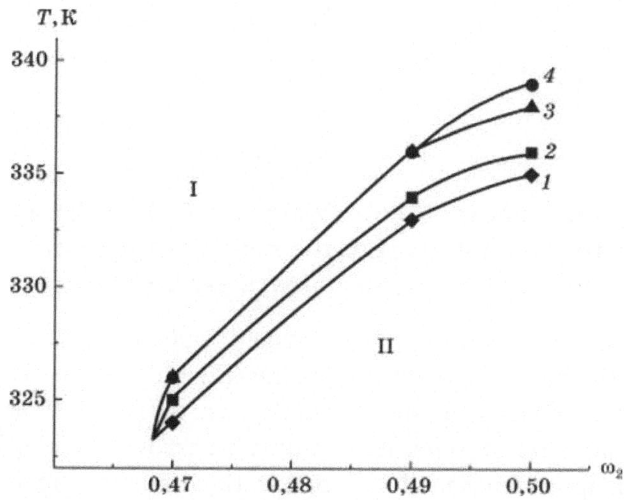

Figure 30. Boundary curves of CEC-DMF system. H = 0 (1), 3 (2), 5 (3), and 9 kOe (4).

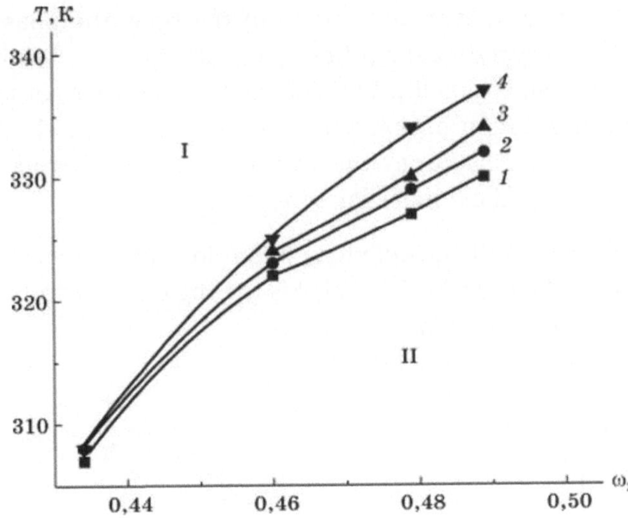

Figure 31. Boundary curves of CEC-DMAc system. H = 0 (1), 3 (2), 5 (3), and 9 kOe (4).

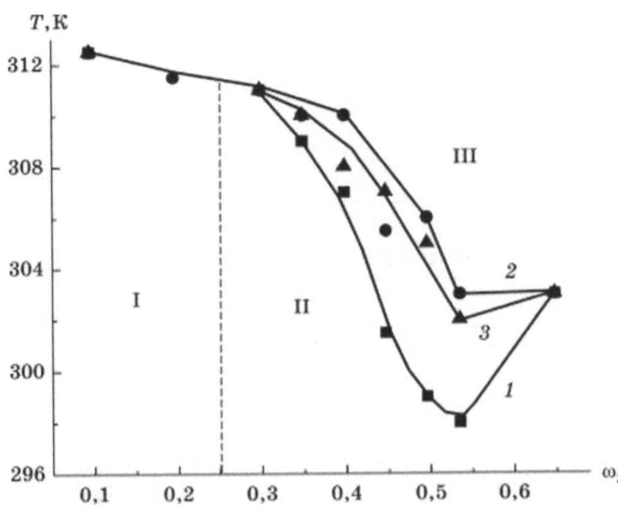

Figure 32. Boundary curves of HPC-1-water. H = 0 (1), 5 (2), and 9 (3) kOe. I is the region of isotropic solutions, II is the region of anisotropic solutions, III is the region of amorphous phase separation.

Figures 29–32 show that the magnetic field leads to the boundary curve shift to the region of the higher temperatures. Thus, the anisotropic (II) region extends and isotropic region (I) gets narrow, and in the case of the HPC-1-water system, the region of amorphous phase separation becomes narrow. The observed effects are related to the change in macromolecule orientation under a magnetic field.

The shift of the boundary curves of the studied systems depends on the intensity of the magnetic field and polymer concentration, as discussed below.

Figure 33 shows the diagrams of ΔT dependence on the magnetic field intensity (H) for the HPC-3-DMAc system, where ΔT is the difference between the temperatures of the phase LC transition in the magnetic field and in its absence [12].

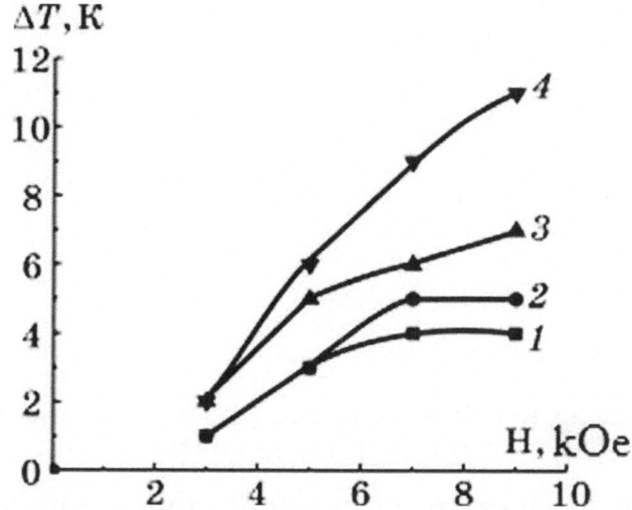

Figure 33. Dependence ΔT vs. H for the HPC-3–DMAc system: ω2 = 0.46 (1), 0.48 (2), 0.50 (3), and 0.51 (4).

The similar behavior is observed for all given systems (except for the HPC-1-water system): the greater the intensity of the magnetic field, the more the macromolecule orientation and the higher the temperature of the LC phase transition. For the HPC-1-water system, the inverse dependence is observed: the higher the value of H, the lower the ΔT. Probably this is associated with the specific character of the solvent and with the other type of the phase transition upon heating.

The effect of the polymer concentration on magnetic-field-induced changes in phase transition temperatures is due to two factors: (1) as concentration increases, the number of macromolecules capable of orientation in the magnetic field grows; as a consequence, the temperature of the phase LC transition should increase; (2) polymer concentration growth leads to the increase in viscosity and in the interchain interaction that hampers the development of the orientation processes and weakens the impact of the magnetic field. In general, the dependence of ΔT on concentration seems to be described by a curve with a maximum similar to the curve of the HPC-1-water system.

It should be noted that for the solutions of the lower-molecular-weight sample HPC, the value of ΔT is significantly larger. This proves the better orientation of smaller molecules in the magnetic field, which coincides with the results of the other studies (Figure 34) [12].

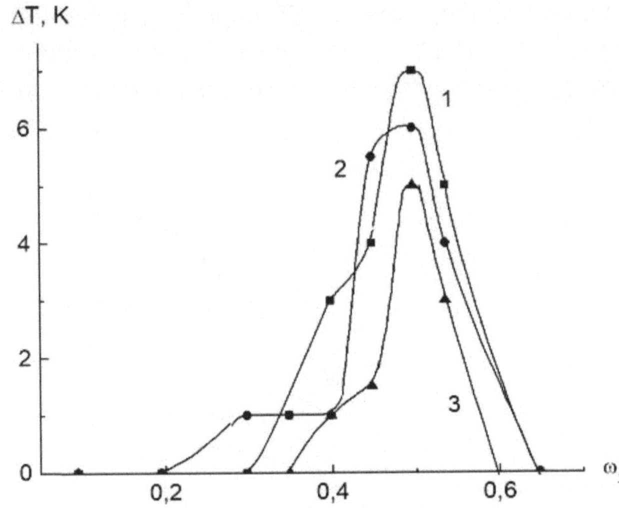

Figure 34. Concentration dependence of ΔT for the HPC-1-water system: H = 5 (1), 9 (2), and 13 kOe (3).

4.1 Magnetic field energy stored by solutions and temperature of phase transitions

The solutions of cellulose derivatives exhibit the formation of the cholesteric liquid crystals [1]. The exposure of solutions in magnetic field induces the transition of the cholesteric liquid crystal to the forced nematic liquid crystal. And such transition has a threshold character with the critical intensity of the magnetic field Hcritical ~ 103 Oe. For the calculation of energy E stored by a unit of the solution volume, the values of magnetic susceptibility χ were determined for a number of the systems. Table 6 lists the results.

Table 6. Magnetic susceptibility of the systems.

System	$-\chi \cdot 10^7$
HPC-1	18.1
HPC-1-DMAc (ω_2 = 0.5)	6.9
HPC-1-water (ω_2 = 0.5)	3.4
HPC-1-CH_3COOH (ω_2 = 0.3)	71.3
CEC	5.3
CEC-DMAc (ω_2 = 0.5)	6.7

All calculated values of χ are negative. Therefore, the studied systems show the diamagnetic properties that agree with their chemical structure. The value of E was calculated according to the equation: E = $-\chi$H2, where H is the intensity of the magnetic field. It was found that the dependence of ΔT vs. lnE is described by the straight lines for the systems: HPC-1-DMAc, HPC-3-DMAc, CEC-DMAc, PBG-DMF (Figures 35 and 36).

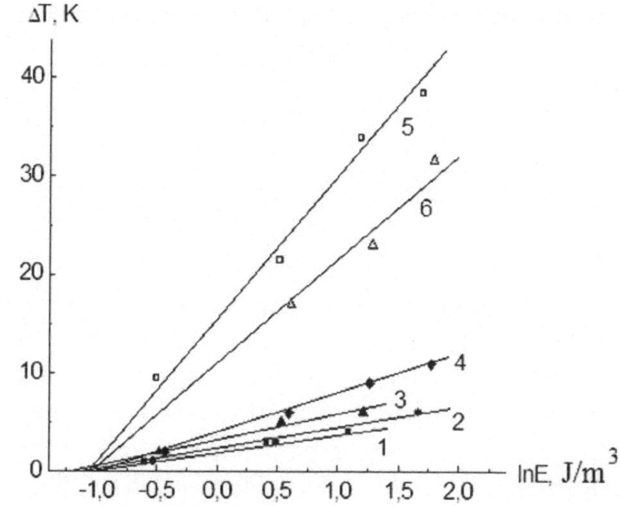

Figure 35. Plot of ΔT vs. lnE for the systems: HPC-3-DMAc: ω2 = 0.46 (1), 0.48 (2), 0.50 (3), and 0.51 (4); HPC-1-DMAc: ω2 = 0.49 (5), and 0.52 (6).

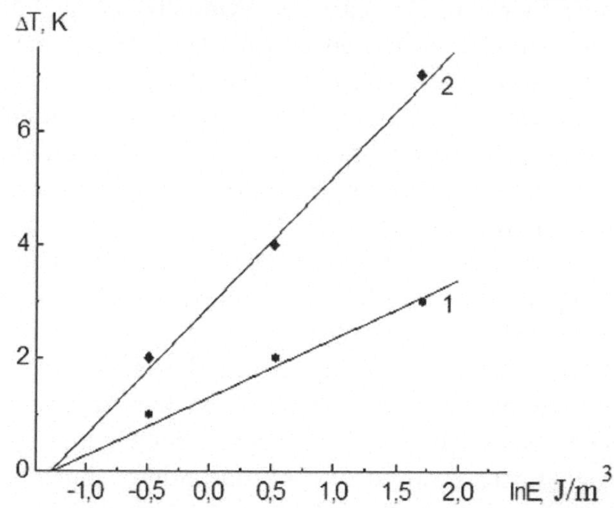

Figure 36. Plot of ΔT vs. lnE for CEC-DMAc system: ω 2 = 0.46 (1) and 0.49 (2).

Figure 35 explicitly shows that the smaller the HPC molecules, the better they are oriented in the magnetic field and the higher the ΔT. Figure 36 also show that the direct dependencies of ΔT on lnE intersect the abscissa axis virtually at one point, i.e., the temperature change of the LC phase formation in solutions begins from a certain value of the energy E0. Thus, the impact of the magnetic field on the phase LC transitions in the solutions of cellulose ethers has a threshold character, i.e., it appears at the magnetic field intensity exceeding the critical value of Hcritical. The computed values of Hcritical given in Table 7 agree with the values given in other studies [1] for Hcritical of the magnetic field required for the conversion of the cholesteric liquid crystal into a nematic one.

Table 7. Calculated values for E0 and Hcritical for systems HPC-DMAc, CEC-DMAc.

System	E_0, J/m^3	$H_{critical}$, kOe
HPC-DMAc	0.34	2.2
CEC-DMAc	0.28	2.0

5. Conclusions

The classification of liquid crystals is considered. The phase diagrams of the polymer-solvent systems with liquid crystalline phase separation are described. It has been shown that, as the molecular mass of cellulose ether increases, the boundary curve separating the isotropic region from the anisotropic one on the phase diagram shifts to the region of higher temperatures and lower concentrations of the polymer. The replacement of the hydroxypropyl radical in cellulose ether units with ethyl or hydroxyethyl radicals leads to a decrease in the concentration of LC phase formation. The introduction of nitro groups into the units of CEC chains increases the rigidity of chains; therefore the degree of asymmetry of macromolecules grows and the LC phase is formed at lower concentrations of the polymer. The higher the polarity of solvent molecules, the more negative the Gibbs energy of mixing and the higher the second virial coefficients, that is, the better the interaction between the components. Thus, in a first approximation, the dipole moment can be used as an estimate of solvent quality for cellulose ethers. The rigid macromolecules orient easily in the external fields. This results in: the type of liquid crystal changes from the cholesteric to the nematic one; the domain structure is formed in the solutions; the concentration dependency of ΔT (ΔT is the difference of Tph in the field and in its absence, respectively) is described by the curve exhibiting maximum. The higher the magnetic field intensity, more evident the effects. The smaller the macromolecules, the more profound are their orientations along the lines of force of magnetic field and the bigger the ΔT value. The ΔT value is determined by the field energy E (E is the energy of a magnetic field, stored by a solution volume unit).

The processes of self-assembly of nanomolecules are most intensive in the solutions of rigid-chain polymers. This process has a “packet” mechanism. In diluted solutions there exist isolated macromolecules, which, within the concentrations of 2–5% (depending on the polymer molecular weight), form supramolecular particles: “packets,” which remain stable (do not change dimensions) in a sufficiently wide range (from 10 to 30%). With a further increase in the polymer concentration, the number of packets (nuclei of the new LC-phase) increases while their sizes are conserved. The system transition to the fully LC state is caused by the aggregation of packets and the formation of large particles.

References

1. Papkov SP, Kulichikhin VG. Liquid Crystalline State of Polymers. Moscow: Chemistry; 1977. 240 p. [in Russian]
2. Reinitzer F. Beitrage zur Kenntniss des Cholesterins. Wiener Monatshefte für Chemie. 1888;9:421-441
3. Lehman O. Uber Fliessende Kristalle. Zeitschrift für Physikalische Chemie. 1889;4:462-472
4. Zhdanov SI, editor. Liquid Crystals. Moscow: Chemistry; 1979. 328 p. [in Russian]
5. Sonin AS. Introduction to Liquid Crystal Physics. Moscow: Nature; 1983. 400 p. [in Russian]
6. Blyumshteyn A, editor. Liquid Crystalline Order in Polymers. Moscow: Mir; 1981. 302 p. [in Russian]
7. Andropov VV, Barmatov EB, Shibaev VP, Philippov AP. Dynamics of orientation of nematic copolymer and its liquid crystal mixtures with low molecular weight additive. Vysokomolekulârnye soedineniâ. Seriâ A. 2002;44:1111-1118
8. Barmatov EB, Medvedev AV, Ivanov SA, Barmatova MV, Shibaev VP. Phase state and photooptical behavior of mixtures of functionalized liquid crystal copolymers with low molecular weight photochromic additives stabilized by hydrogen bonds. Vysokomolekulârnye soedineniâ. Seriâ A. 2001;43:468-477
9. Kimura T. Study on the effect of magnetic field on polymeric materials and its application. Polymer Journal. 2003;35:823-843. DOI: 10.1295/polymj.35.823
10. Moore JS, Stupp SI. Orientation dynamics of main-chain liquid crystal polymers. 2. Structure and kinetics in a magnetic field. Macromolecules. 1987;20:282-293. DOI: 10.1021/ma00168a009
11. Yamato M, Tsunehisa Kimura T. Magnetic Processing of Diamagnetic Materials. Polymers. 2020;12:1491-1514. DOI: 10.3390/polym12071491
12. Vshivkov SA. Phase Transitions and Structure of Polymer Systems in External Fields. Newcastle: Cambridge Scholars Publishing; 2019. 370 p
13. Vshivkov SA, Rusinova EV. Magnetorheology of polymer systems. In: Rivera-Armenta JL, Salazar Crus BA, editors. Polymer Rheology. London: InTechOpen; 2018. pp. 3-28. DOI: 10.5772/intechopen.75768
14. Meuer RB. Distortion of a cholesteric structure by a magnetic field. Applied Physics Letters. 1969;14:208-212. DOI: 10.1063/1.1652780
15. de Gennes PG. Calcul de la distorsion d"une structure cholesterique par un champ magnetique. Solid State Communications. 1968;6:163-165. DOI: 10.1016/0038-1098(68)90024-0
16. DuPre DB, Duke RW. Temperature, concentration, and molecular weight dependence of the twist elastic constant of cholesteric poly-γ-benzyl-L-glutamate. The Journal of Chemical Physics. 1975;63:143-148. DOI: 10.1063/1.431066

17. Duke RW, DB DP, Hines WA, Samulski ET. Poly(γ-benzyl L-glutamate) helix-coil transition. Pretransition phenomena in the liquid crystal phase. Journal of the American Chemical Society. 1976;98:3094-3101

18. Miller WG, Wu CC, Wee EL, Santee GL, Rai JH, KD, Goebel KD. Thermodynamics and dynamics of polypeptide liquid crystals. Pure and Applied Chemistry. 1974;38:37-58. DOI: 10.1351/pac197438010037

19. Ciambella J, Stanier DC, Rahatekar SS. Magnetic alignment of short carbon fibres in curing composites. Composites. Part B, Engineering. 2017;109:129-137. DOI: 10.1016/j.compositesb.2016.10.038

20. Teranishi S, Kusumi R, Kimura F, Kimura T, Aburaya K, Maeyama M. Biaxial magnetic orientation of zinc citrate as nucleating agent of poly(L-lactic acid). Chemistry Letters. 2017;46:830-832. DOI: 10.1246/cl.170092

21. Horii S, Arimoto I, Doi T. Linear drive type of modulated rotating magnetic field for a continuous process of three-dimensional crystal orientation. Journal of the Ceramic Society of Japan. 2018;126:885-888. DOI: 10.2109/jcersj2.18128

22. Chung JY, Lee JG, Baek YK, Shin PW, Kim YK. Magnetic field-induced enhancement of thermal conductivities in polymer composites by linear clustering of spherical particles. Composites. Part B, Engineering. 2018;136:215-221. DOI: 10.1016/j.compositesb.2017.10.033

23. Vshivkov SA, Avvakumova AS. Effect of magnetic field on the rheological properties of poly(ethylene glycol) and poly(dimethylsiloxane) mixtures with aerosil and iron nanoparticles. Polymer Science, Series A. 2017;59:764-771. DOI: 10.1134/S0965545X17050170

24. Vshivkov SA, Zhernov IV, Nadol'skii AL, Mizyov AS. Effect of magnetic field on phase transitions in solutions and melts of flexible polymers. Polymer Science, Series A. 2017;59:465-472. DOI: 10.1134/S0965545X17040149

25. Naga N, Ishikawa G, Noguchi K, Takahashi K, Watanabe K, Yamato M. Magnetic-field induced alignment of low molecular weight polyethylene. Polymer (Guildf). 2013;54:784-790. DOI: 10.1016/j.polymer.2012.12.016

26. Vshivkov SA, Rusinova EV, Klyuzhin ES, Kapitanov AA. Effect of magnetic field on phase transitions and structure of polyelectrolyte solutions. Polymer Science. A. 2020;62:62-69. DOI: 10.1134/S0965545X19050183

27. Naga N, Saito Y, Noguchi K, Takahashi K, Watanabe K, Yamato M. Magnetic-field-induced alignment of syndiotactic polystyrene. Polymer Journal. 2016;48:709-714. DOI: 10.1038/pj.2016.21

28. Gopinadhan M, Choo Y, Osuji CO. Strong orientational coupling of block copolymer microdomains to smectic layering revealed by magnetic field alignment. ACS Macro Letters. 2016;5:292-296. DOI: 10.1021/acsmacrolett.5b00924

29. Vshivkov SA, Rusinova EV, Kudrevatykh NV, Galyas AG, Alekseeva MS, Kuznetsov DK. Phase transitions of hydroxypropylcellulose liquid-crystalline solutions in magnetic field. Polymer Science. A. 2006;48:1115-1119. DOI: 10.1134/S0965545X06100142

30. Vshivkov SA, Rusinova EV, Kutsenko LI, Galyas AG. Phase transitions in liquid-crystalline cyanoethyl cellulose solutions in magnetic field. Polymer Science. B. 2007;49:200-202. DOI: 10.1134/S156009040707007X

31. Vshivkov SA, Rusinova EV. Effect of magnetic fields oh phase transitions in solutions of cellulose derivatives. Polymer Science. A. 2008;50:725-732. DOI: 10.1134/S0965545X08070018

32. Vshivkov SA, Rusinova EV, Galyas AG. Phase diagrams and rheological properties of cellulose ether solutions in magnetic field. European Polymer Journal. 2014;59:326-332. DOI: 10.1016/j.eurpolymj.2014.07.042

33. Vshivkov SA, Rusinova EV, Galyas AG. Effect of a magnetic field on the rheological properties of the rheological properties of cellulose ether solutions. Polymer Science. A. 2012;54:827-832. DOI: 10.1134/S0965545X1211003X

34. Vshivkov SA, Rusinova EV, Galyas AG. Relaxation character of the rheological behavior of cellulose ether solutions in a magnetic field. Russian Journal of Applied Chemistry. 2014;87:1140-1145. DOI: 10.1134/S1070427214080217

35. Vshivkov SA, Soliman TS. Effect of a magnetic field on the rheological properties of the systems hydroxypropyl cellulose – ethanol and hydroxypropyl cellulose – dimethyl sulfoxide. Polymer Science. A. 2016;58:307-314. DOI: 10.1134/S0965545X16030184

36. Vshivkov SA, Soliman TS. Phase transitions, structures, and rheological properties of hydroxypropyl cellulose—ethylene glycol and ethyl cellulose—dimethylformamide systems in the presence and in the absence of a magnetic field. Polymer Science. A. 2016;58:499-505. DOI: 10.1134/S0965545X16040143

37. Plate NA, editor. Liquid-crystalline polymers. Moscow: Chemistry; 1988. 416 p. [in Russian]

38. Onsager L. The effect of shape on the interaction of colloidal particles. Annals of the New York Academy of Sciences. 1949;51:627-659

39. Flory PJ. Phase equilibria in solutions of rod-like particles. Proceedings of the Royal Society of London A. 1956;234:73-89

40. Miller WG, Wu CC, Wee EL, Santee GL, Rai JH, Gaebel KG. Thermodynamics and dynamics of polypeptide liquid crystals. Pure and Applied Chemistry. 1974;38:37-54. DOI: 10.1351/pac197438010037

41. Iovleva MM, Papkov SP, Mil'kova LP, Kalmykova VD, Volokhina AV, Kudryavtsev GI. Temperature-concentration boundaries of liquid crystalline state of poly-p-benzamide. Polymer Science. U.S.S.R. B. 1976;18:830-832

42. Papkov SP, Kulichikhin VG, Kalmykova VD, Malkin AY. Journal of Polymer Science Polymer Physics Edition. 1974;12:1753-1770

43. Sasaki S, Tokuma K, Uematsu I. Phase behavior of poly(γ-benzyl L-glutamate) solutions in benzyl alcohol. Polymer Bulletin. 1983;10:539-546. DOI: 10.1007/BF00285373

44. Nakajima A, Hirai T, Hayashi T. Phase relationship of rodlike polymer, poly(p-phenyleneterephthalamide) in sulfuric acid-water system. Polymer Bulletin. 1978;1:143-147. DOI: 10.1007/BF00255729
45. Kiss G, Porter RS. Rheology of the concentrated solutions of poly-γ-benzyl-L-glutamate. American Chemical Society, Polymer Preprints. 1977;18:185-186
46. Lukashova NV, Volokhina AV, Papkov SP. Phase conversions and viscosity properties of aromatic copolyamide solutions. Vysokomolekulârnye Soedineniâ. Seriâ B. 1978;20:151-154
47. Konevets VI, Andreeva VM, Tager AA, Ershova IA, Kolesnikova EN. Studying of the moderate-concentrated solution structure of some polyamides in the composition region, preceding the liquid crystal formation. Vysokomolekulârnye soedineniâ. Seriâ A. 1985;27:959-967
48. Kulichikhin VG, Golova LK. Liquid-crystalline state of cellulose and cellulose derivatives. Khim Drev. 1985;3:9-27
49. Zugenmaier P, Vogt U. Structural investigations on cellulose tricarbanilate: Conformation and liquid crystalline behavior. Macromolekulare Chemie. 1983;194:1749-1760. DOI: 10.1002/macp.1983.021840818
50. Werbowyj RS, Gray DG. Ordered phase formation in concentrated hydroxpropylcellulose solutions. Macromolecules. 1980;13:69-73
51. Conio G, Bianchi E, Ciferri A, Tealdi A, Aden MA. Mesophase formation and chain rigidity in cellulose and derivatives: 1. (Hydroxypropyl) cellulose in dimethylacetamide. Macromolecules. 1983;16:1264-1270
52. Vshivkov SA, Rusinova EV. Effect of component nature on liquid – crystalline transitions in solutions of cellulose ethers. Polymer Science. A. 2018;60:65-73. DOI: 10.1134/S0965545X18010078
53. Vshivkov SA, Galyas AG, Kutsenko LI, Tyukova IS, Terziyan TV, Shepetun AV. Self-organization of macromolecules and liquid-crystalline phase transitions in solutions of cellulose ethers. Polymer Science. A. 2011;53:1-5. DOI: 10.1134/S0965545X1101010X
54. Vshivkov SA, Galyas AG. Mechanism of self-assembly of rigid–chain macromolecules of cellulose ethers in solutions. Polymer Science. A. 2011;53:1032-1039. DOI: 10.1134/S0965545X11100130
55. Vshivkov SA, Adamova LV, Lirova BI. Thermodynamic parameters of the interaction of cellulose ethers with low-molecular-mass liquids. Polymer Science. A. 2012;54:821-826. DOI: 10.1134/S0965545X12080123
56. Shimamura K, White JL, Feller JF. Hydroxypropylcellulose, a thermotropic liquid crystal: Characteristics and structure development in continuous extrusion and melt spinning. Journal of Applied Polymer Science. 1981;26:2165-2180. DOI: 10.1002/app.1981.070260705
57. Charlet G, Gray DG. Solid cholesteric films cast from aqueous (hydroxypropyl) cellulose. Macromolecules. 1987;20:33-38. DOI: 10.1021/ma00167a007

58. Fischer H, Murray M, Keller A, Odell JA. On the phase diagram of the system hydroxypropylcellulose-water. Journal of Materials Science. 1995;30:4623-4627. DOI: 10.1007/BF01153071

59. Belousov VP, Panov MY. Thermodynamics of Nonelectrolyte Aqueous Solutions. Leningrad: Khimiya; 1983. 264 p. [in Russian]

60. Vshivkov SA, Adamova LV, Rusinova EV, Safronov AP, Dreval VE, Galyas AG. Thermodynamics of liquid-crystalline solutions of hydroxypropyl cellulose in water and ethanol. Polymer Science. A. 2007;49:578-583. DOI: 10.1134/S0965545X07050124

61. Fortin S, Charlet G. Phase diagram of aqueous solutions of (hydroxypropyl) cellulose. Macromolecules. 1989;22:2286-2292. DOI: 10.1021/ma00195a050

62. Kamide K, Saito M, Abe T. Dilute solution properties of water-soluble incompletely substituted cellulose acetate. Polymer Journal. 1981;13:421-431. DOI: 10.1295/polymj.13.421

63. Ravdel AA, Ponomareva AM. Short Handbook of Physicochemical Values. Leningrad: Chemistry. 232 p

64. Tager AA. Foundations of Nonelectrolyte Solutions Science. Yekaterinburg: Ural. Gos. University; 1993. 312 p. [in Russian]

65. Heller W, Bhathagar HL, Nakagaki M. Theoretical investigations on the light scattering of spheres. XIII. The "wavelength exponent" of differential turbidity spectra. The Journal of Chemical Physics. 1962;36:1163-1170. DOI: 10.1063/1.1732710

66. Heller W, Pangonis WL. Theoretical investigations on the light scattering of colloidal spheres. I. The specific turbidity. The Journal of Chemical Physics. 1957;26:498-505. DOI: 10.1063/1.1743332

67. Heller W. Theoretical investigations on the light scattering of spheres. XV. The wavelength exponents at small α values. The Journal of Chemical Physics. 1964;40:2700-2704. DOI: 10.1063/1.1725583

68. Klenin VI, Shchegolev SY, Lavrushin VL. Characteristic Functions of Light Scattering from Disperse Systems. Saratov: Saratovsk. Gos. University; 1977. 177 p. [in Russian]

69. Schulz L, Seger B, Buchard W. Structures of cellulose in solution. Macromolecular Chemistry and Physics. 2000;201:2008-2022. DOI: 10.1002/1521-3935(20001001)201:15<2008::AID-MACP2008>3.0.CO;2-H

70. Kuhn W, Moser P. Eifluss der Bildung loser Molekulaggregate auf die Viskositat der Losungen macromolekular Stoffe. Macromolekulare Chemie. 1961;44-46:71-77. DOI: 10.1002/macp.1961.020440107

4

Molecular Simulation of Cholesteric Liquid-Crystal Polyesteramides: Conformational and Structure Analysis by Rietveld Refinement

1. Introduction

In 1953, K. Ziegler discovered a catalytic system to obtain industrially linear crystalline polyethylene at mild temperatures and pressures (based on titanium halides and organoaluminium compounds). In 1954, G. Natta developed crystalline polypropylene and proved the unprecedented regularity of its molecules and helical form of its crystals. In 1963 both scientists shared the Nobel Prize in Chemistry by having created the science of "regular polymers and stereospecific polymerization" [1].

Wallace Carothers had stated in 1929, working at the DuPont company, the aim to prepare polymers of particular structure through the use of established organic reactions and he had demonstrated the relationship between structure and properties for some polymers: polyesters, polyamides, neoprene, etc. He had first claimed that structural and stereochemical factors will usually be more important factors than others, such as temperature, in bifunctional condensation reactions, explained in terms of a step-growth of bi-functional monomers reacting to first form dimers, then trimers, longer oligomers and eventually long-chain polymers [2].

The problem arose of the relationship between the conformation of stereoregular macromolecules in the crystalline state and in solution, as well as between stereoregularity and conformation. The research in this field stimulated the synthesis of optically active polymers, particularly vinyl polymers, with the hope that optical activity might be of help in investigating the conformation of macromolecules in solution [3].

Since then, the synthesis and application of optically active synthetic polymers and their structure control have attracted growing interest. Three major groupings can be established among them: polymers with optical activity arising from asymmetric centers in the side or main chain; polymers with optical activity arising from both asymmetric centers and macromolecular asymmetry based on the secondary structure (i.e., a helix); and polymers with optical activity arising entirely from conformational asymmetry (macromolecular asymmetry) when the polymers are prepared by helix-sense selective polymerization [4].

Chirality is essential for life and chiral phenomena play significant roles in nature.

Most of the naturally occurring molecules/macromolecules, such as nucleic acids, proteins, and polysaccharides are chiral and optically active. This situation can be very obviously seen if we look at the chirality of nearly 800 drugs (about 97%) derived from natural sources. Only 2% are racemates and only 1% is achiral [5].

The parallel behavior between liquid-crystals in materials science and lipids in life science was pointed out by Ringsdorf in 1988, based on their common amphiphilic nature, being both self-organizing systems [6].

In 1992 the International Union of Crystallography re-defined the concept of crystal as: "Any solid which has a diffraction pattern essentially discrete" (Fourier space) [7]. The Crystal Family was accepted to be composed of: Periodic and aperiodic crystals. Liquid-crystals (LC) belonging to the last group.

DNA behaves as a liquid-crystal with a cholesteric mesophase, described as the special array of nematic planes, containing complementary base pairs, stacked in a superstructure with chiral helical symmetry of charge distribution [8–10].

Two polyesteramides designed as PNOBDME (C34H38N2O6)n and PNOBEE (C26H22N2O6)n in Figure 1a and b, have been synthesized and characterized as cholesteric liquid crystals -through the condensation reaction between 4 and 4'-(terephthaloyl- diaminedibenzoic chloride (NOBC) and racemic glycol: DL-1,2 dodecanediol, or DL-1,2-butanediol, respectively [11–13], being chemical modifications of precursor multifunctional cholesteric LC polyesters PTOBDME [C34H36O8]n and PTOBEE [C26 H20 O8]n [14–16], shown in Figure 2, with newly added properties but holding their helical macromolecular structures. All of them have been shown to be biocompatible against macrophages and fibroblasts cellular lines and behave as both thermotropic and lyotropic. Besides, they can interact with biomacromolecules such as lipids (both neutral and cationic) and with polynucleotides and nucleic acids. They have proved to be able to act as non-viral vectors in gene therapy [17, 18].

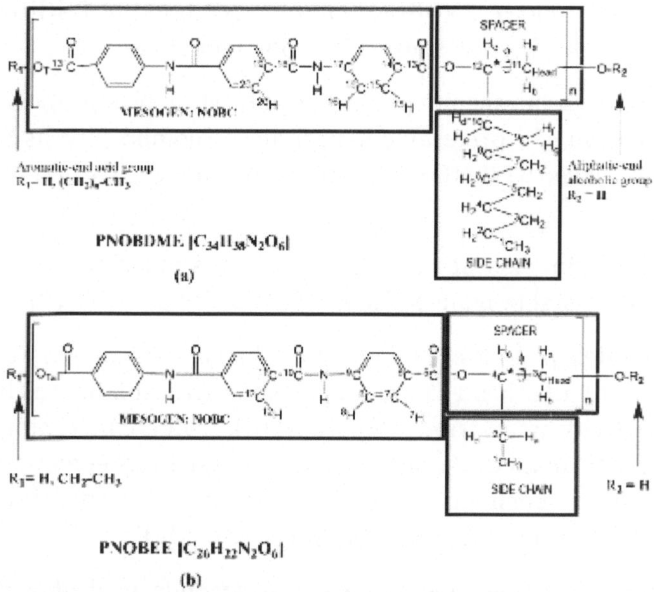

Figure 1. The monomeric unit of polyesteramides: (a) PNOBDME; (b) PNOBEE. The three different zones of the monomer: mesogen, spacer and flexible side chain are indicated. The asterisk indicates the chiral center (12C*) in PNOBDME and (4C*) in PNOBEE, both in the flexible spacer. Torsion angles φ, along with the 11C-12C* bond and 3C-4C* bond, respectively, are indicated. Aromatic-end acid and aliphatic-end alcoholic groups are also specified.

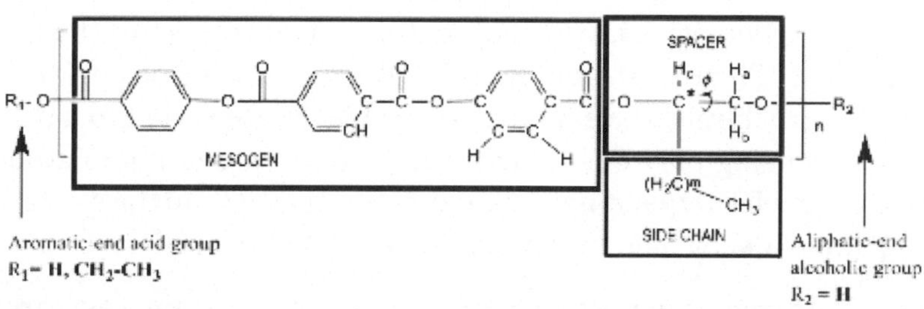

Figure 2. Monomeric unit of precursor cholesteric liquid-crystalline polyesters PTOBDME (m = 9) and PTOBEE (m = 1). The asterisk in the spacer indicates the chiral center (12C*) in PTOBDME and (4C*) in PTOBEE.

Although the starting raw materials were all racemic, the studied cholesteric LC polymers exhibit unexpected optical activity and chiral morphology.

According to our previous experience, each enantiomeric polymer shows two independent sets of signals, observed by conventional NMR techniques, attributed to two diastereomeric conformers: gg and gt, differentiated with the apostrophe (')

and without it (). They have been related with two possible staggered conformations of the torsion angle φ- containing the asymmetric carbon atom in the spacer- along the 11C-12C* bond in PNOBDME and 3C-4C* bond in PNOBEE respectively, with two possible helical screw sense of the polymer chain.

Chirality observed in racemic precursor polyesters was proposed to be due to the kinetic resolution of a preferable helical diastereomer, such as Sgt, concerning the possible four forms, while the R/S ratio of asymmetric carbon atoms remained 50/50.

The same behavior has been observed in six similar functionalized cationic cholesteric liquid crystal polymers also synthesized in our lab [19].

Molecular models of all these LC polymers had always shown helical polymeric chains with stereoregular head-tail, isotactic structure, explained as being the result of the higher reactivity of the primary hydroxyl OH-groups concerning the secondary ones, in the glycol, through the condensation reaction. In agreement with Carothers, the structural and stereochemical factors are crucial elements in the progress of polymer structure through the condensation reaction [2].

The structural fragment, including chiral secondary alcohol and a primary alcohol group (a beta-chiral 1,2 diol) is particularly interesting since it is present in many relevant natural products, such as sugars, nucleosides, glycerides) [20], chiral nanostructures from helical polymers and metallic salts [21].

A detail of the spacer is visible in Figure 3, showing the secondary alcohol group bonded to chiral 12C* in PNOBDME (to 4C* in PNOBEE) and the primary alcohol bonded to 11C (to 3C in PNOBEE).

Tetrahedral carbon atoms 11C, allocated in α with respect to the asymmetric carbon atom 12C* (Figure 1a), and 3C in PNOBEE, in α with respect to chiral 4C* (Figure 1b), are prochirals, since both could be ideally converted to a chiral centre by arbitrarily changing only one attached H group to a deuterium atom (D with higher priority than H). Depending on the configuration, R/S, of the so created chiral centre, the H atom ideally deuterated, would be labeled as pro-R/S.

The two hydrogen atoms Ha and Hb, bonded to prochiral 11C carbon atom, in PNOBDME, can be described as prochiral hydrogens, also designated as diastereotopic. Their indistinguishable signals by 1H-NMR split then into two signals easily differentiated. The same effect is observed for Hd and He, bonded to prochiral 10C, and for Hf and Hg, bonded to prochiral 9C. Equally the two hydrogen atoms on prochiral 3C carbon atom Ha and Hb, and Hd, He bonded to 2C in PNOBEE, are considered as prochiral hydrogens.

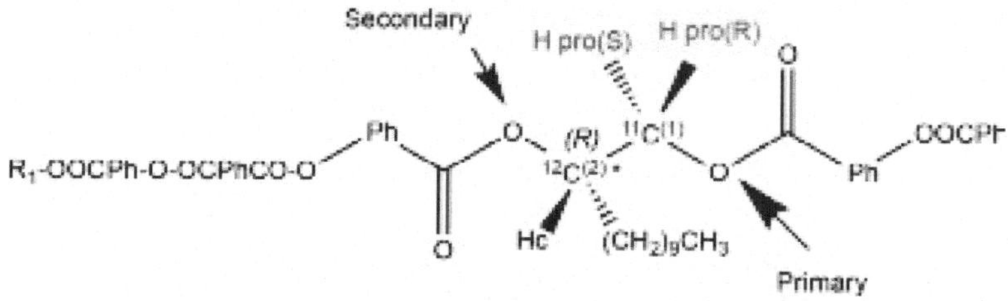

Figure 3. Scheme of the spacer in PNOBDME including the two alcohol groups with R absolute configuration of 12C*.

Besides chirality due to the asymmetrical carbon atom, a second chirality due to the helical conformation is predicted by the molecular models of polyesteramides PNOBDME and PNOBEE.

According to Ute et al. [22–24], when an isotactic polymer with asymmetric carbon atoms carries non-identical end-groups (R1 R2), each polymer can have two enantiomers, e.g., R1-RRR R-R2 and R1-SSS S-R2 isomers. Each enantiomer with two helical conformations, right-handed and left-handed, being diastereomeric structures whose energies differ from one another (i. e. dG 0). Each helical state can be observed by NMR as an independent set of signals, while enantiomers are not distinguishable by NMR.

In the case of PNOBDME, the presence of the chiral carbon atom 12C* in the repeat unit and the different end groups for the aliphatic-end (-OH) and aromatic-end (-COOH) promotes the presence of two helical conformations gg and gt, experimentally observed in the NMR spectra as two independent sets of signals differentiated as (´) and () for each isotactic stereoregularity R, R, R, and S, S, S, that implies four possible diastereoisomers. Figure 4 shows the four helical conformers of PNOBDME and PNOBEE through the 11C-12C* bond and 3C-4C* (torsion φ), respectively.

The existence of these two independent conformers had also been observed for PTOBDME and PTOBEE by Raman experiments [25] and by NMR [26]. The combination of a helix with two screw senses and the two absolute configurations providing four diastereomeric structures. It was also related to the presence of helical structures, the Cotton effect and the sign of the helicity in the case of 1–2 di-O-benzoylated sn-glycerols [27–29].

Details of molecular models for gg and gt conformers of a dimer of PNOBDME are shown in Figure 5, projected along with the 11C-12C* bond, torsion φ, (perpendicular to the paper) with 12C* (bonded to Hc) having R and S absolute configuration, in yellow, behind 11C (bonded to Ha and Hb).

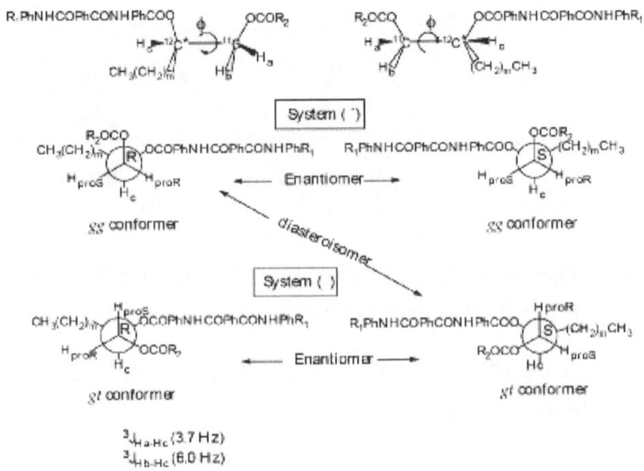

Figure 4. The relationship between the four helical conformations gg and gt of PNOBDME, and PNOBEE through the 11C-12C* bond, and 3C-4C* (torsion φ), respectively.

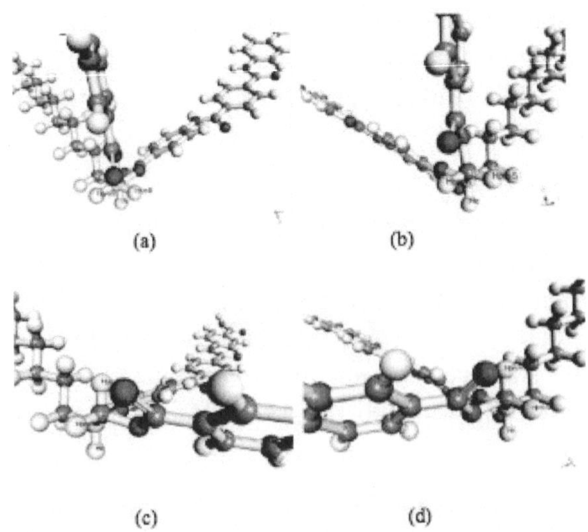

Figure 5. Molecular model details of a PNOBDME dimer. View along 11C-12C* bond (perpendicular to the paper), with (R) and (S) absolute configuration of 12C* (in yellow behind 11C) for: (a) Rgg-diastereoisomer; (b) Sgg-diastereoisomer; (c) Rgt- diastereoisomer; (d) Sgt diastereoisomer.

2. Molecular mechanics simulation and conformational analysis of PNOBEE and PNOBDME

Molecular Mechanics modeling was performed with Materials Studio® v.2021 [30]. COMPASS-III forcefield was used for the monomer, assigning both atomic masses and partial charges. The total energy was then minimized and geometry optimized with FORCITE module.

Within the monomer repeat unit, a Head (CHead) and a Tail (OTail) atoms were defined. The extended length (l) of the bonds in the chain between CH and OT, backbone atoms, including double bonds and rings, is ≈ 23 Å, following the notation by Flory of the spatial representation of a single bonded carbon chain skeleton [31], Figure 6.

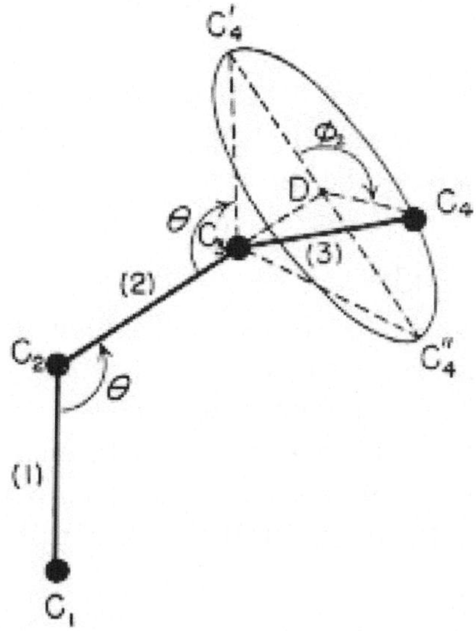

Figure 6. Spatial representation of a single bonded carbon chain after Flory [31].

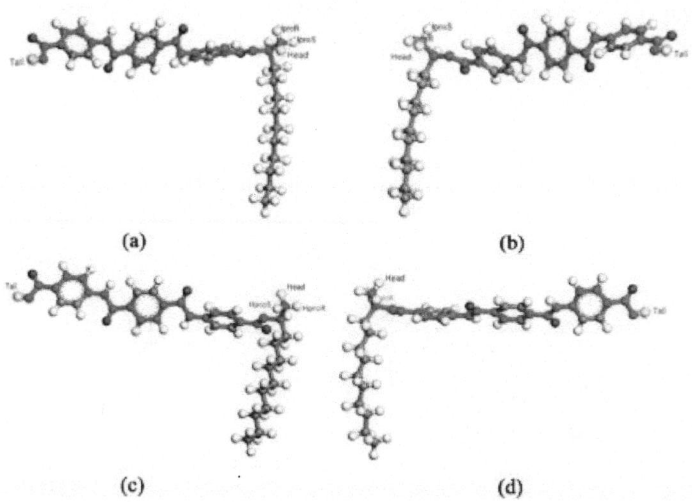

Figure 7. Monomeric repeat units of PNOBDME (a) Rgg-diastereoisomer; (b) Sgg- diastereoisomer; (c) Rgt- diastereoisomer; (d) Sgt diastereoisomer.

According to Flory, the configuration of a chain molecule in space is determined by the lengths l of the bonds of the chain skeleton, by the angles θ specifying the difference between the directions of successive bonds, and by the angles φ of rotation, or torsion, about these bonds. It is possible to consider the lengths of the bonds in the chain and the angle between successive bonds to be fixed quantities for a given structure.

The presence of the chiral centre (C*) atom was considered, including both R or S chirality.

The monomeric repeat units of PNOBDME, are modelized in Figure 7, for the four helical conformations: (a) Rgg; (b) Sgg; (c) Rgt; (d) Sgt, with the defined Head atom (CHead) selected in blue and the Tail atom (OTail) selected in red, and the backbone atoms colored in pink. HproS and HproR are also labeled.

Since the mesogen of PNOBDME and PNOBEE consists of three planar rigid groups: one terephthalamide and two benzoates, limited by parenthesis in Figure 8, connected by amide and ester groups, a simplified model could be reduced to the three articulated planes I, II and III, in the molecular scheme of Figure 9, connected by torsions CN1, N1CB between Plane I and Plane II, and N2C, CBN2 between Plane II and Plane III. A similar simplified model has been defined for the monomeric repeat unit of precursor polyesters PTOBDME and PTOBEE, in Figure 10 [25].

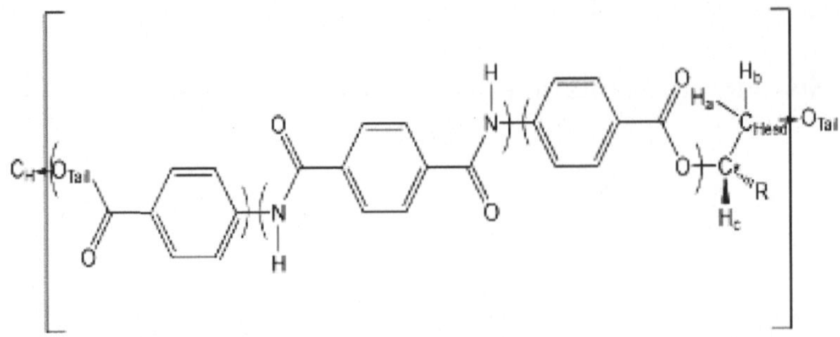

Figure 8. Monomeric repeat unit of PNOBDME and PNOBEE.

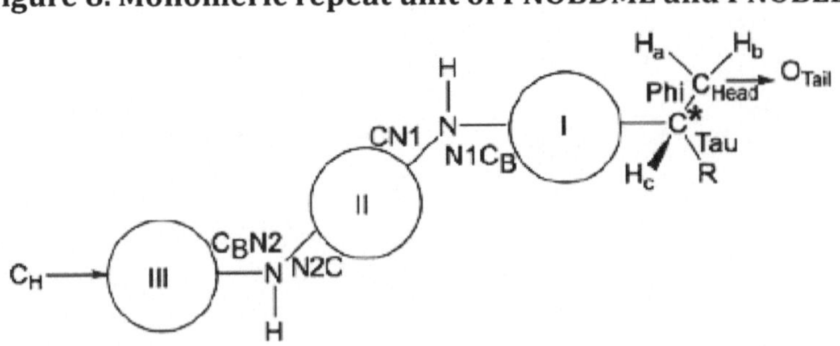

Figure 9. A simplified model of the monomeric repeat unit of PNOBDME and PNOBEE.

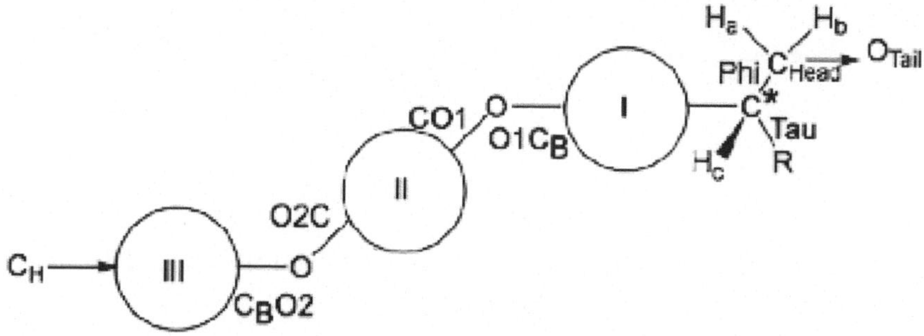

Figure 10. A simplified model of the monomeric repeat unit of precursor polyesters PTOBDME and PTOBEE.

Four more independent torsions were considered within the flexible spacer as Phi (φ), which defines the head orientation to the mesogen, Tau which determines the lateral chain orientation to the mesogen, the torsion CO along with the CcarbonylOester bond (i.e. 13C in PNOBDME and 5C in PNOBEE), close to 180°, and C*-CHead. These torsions were the independent variables for Conformational Analysis (CA). Similar notations has been considered in precursor polyesters PTOBDME and PTOBEE, Figure 10.

1.1 Conformational analysis of PNOBEE and PTOBEE

The molecular model of the monomeric repeat unit of polyesteramide PNOBEE, with a shorter lateral chain, can be seen in Figure 11a after its energy was minimized by performing a geometry optimization with the Forcite module of Materials Studio® [30], compared to that of monomeric precursor polyester PTOBEE, with also a shorter lateral chain, in Figure 11b.

The independent torsion values of PNOBEE monomer, are given in Table 1.

They can be compared to those of PTOBEE monomer in Table 2. PNOBEE exhibits higher CN1 and N2C absolute values than PTOBEE (CO1 and O2C) but lower N1CB and CBN2 absolute values than O1CB and CBO2.

Conformers of PNOBEE with torsions N1CB and CBN2 with the same sign, that means, with two successive rotations of the benzene groups in the same direction, along the main chain axis (i. e. between planes I, II and III): around +35°; +35°

or -35°; -35°, promote the formation of a helix with a triangular cross-section (marked with Δ in Table 1), either with S or R chirality. This is also observed in PTOBEE with O1CB and CBO2: +44°; +44° or 44°; 44°, also signaled with a triangular cross-section (Δ) in Table 2.

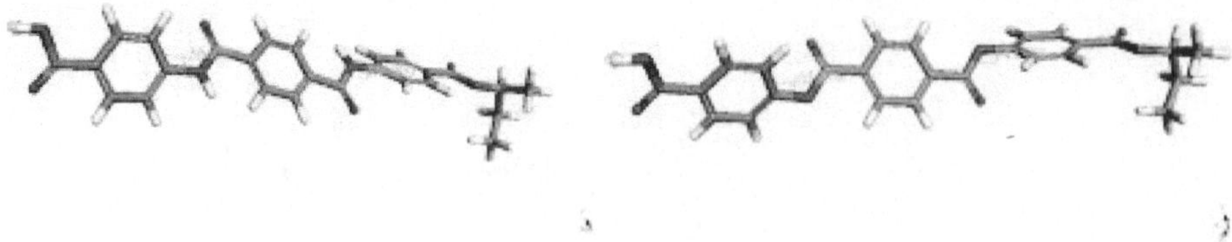

(a) (b)

Figure 11. Molecular models of the optimized monomeric PNOBEE (a) and PTOBEE (b).

Table 1. Conformational analysis of PNOBEE monomer. Torsional angles (°) for the possible conformers.

PNOBEE				Mesogen				Spacer		
		Section	CN1	N1CB °	N2C	CBN2	CO	Phi (φ)	Tau	C*-CH
PNOBEE-R	△		179,2	-35	-180	-36	-180	-82	-54	154,7
-PNOBEE-R	△		-180	35	-180	34	-180	-82,5	-53,5	154,3
PNOBEE-S	△		180	-34	180	-36	179,4	82,7	53,5	-154
- PNOBEE-S	△		-179,1	34	179,1	34	180	81,7	54	-155
PNOBEE-R //	≠		179,1	-36,4	-179	36,3	-179,4	-82	-54	155
-PNOBEE-R //	≠		-179,2	35,4	-179,3	-34,7	-180	-82,2	-54	155
PNOBEE-S //	≠		180	-36,4	-179,1	35,7	179,1	82,1	53,5	-155
-PNOBEE-S //	≠		-180	35,2	180	-34,2	179,5	81,5	53,6	-155,4

Table 2. Conformational analysis of PTOBEE monomer. Torsional angles (°) for the possible conformers.

PNOBEE				Mesogen				Spacer		
		Section	CO1	O1CB	O2C	CBO2	CO	Phi (φ)	Tau	C*-C_H
PNOBEE-R	△		163,2	-44	163,25	-44	172,6	-79,9	-51,9	163,8
-PNOBEE-R	△		-163,5	44	-163,25	44	172,6	-79,9	-51,9	163,8
PNOBEE-S	△		163,3	-44	163,4	-44	-172,5	72,8	51,9	-163,8
- PNOBEE-S	△		-163,3	44	-163,4	44	-172,5	72,8	51,9	-163,8

PNOBEE					Mesogen				Spacer	
		Section	CO1	O1CB	O2C	CBO2	CO	Phi (φ)	Tau	C*-C_H
PNOBEE-R //	≠		163,4	-44	-163,4	44	172,5	-72,9	-51,8	163,8
-PNOBEE-R //	≠		-163,4	44	163,3	-44	172,5	-72,8	-51,9	163,2
PNOBEE-S //	≠		163,4	-44	-163,3	44	-172,5	72,9	51,9	-163,8
-PNOBEE-S //	≠		-163,4	44	-163,3	-44	-172,5	72,9	51,9	-163,8

In conformers of PNOBEE with two successive rotations of the benzene groups in the opposite direction, means torsions N1CB and CBN2 with opposite sign: 35°; +35°; or +35°; 35°, promotes a helix with planes I and III parallel along with the backbone atoms -cross-view- (section in Table 1) for both S or R chirality. They

have been differentiated with //. This is also observed in PTOBEE with O1CB and CBO2 values: 44°; +44°; or + 44°; 44° in Table 2.

The two successive rotations of the benzene groups within the monomer imply the formation of helical molecules, involving a second order of chirality, even without presence of a chiral atom. In Figures 12 and 13 the respective sectional views of the monomeric conformers of PNOBEE and PTOBEE are represented along with the backbone atoms with the three benzene groups forming a triangular cross-section (a); and with Planes I and III parallel (b), for both R and S chirality.

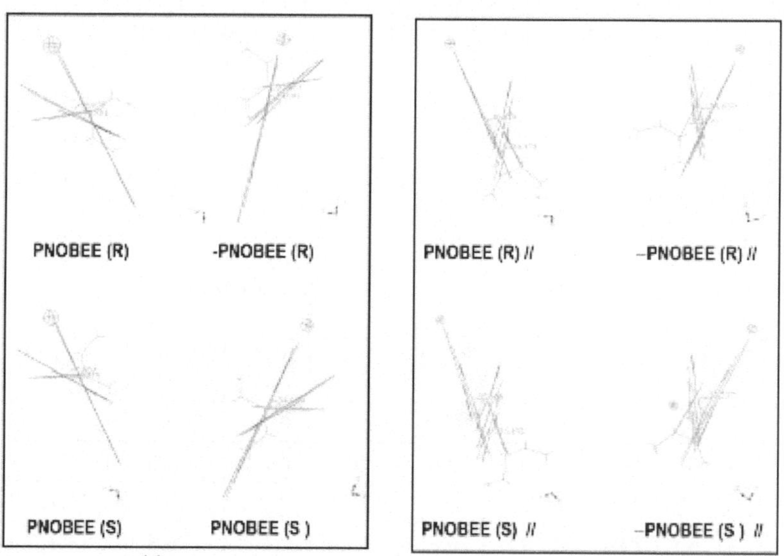

Figure 12. Conformers of monomeric PNOBEE with both R and S chirality, with a triangular sectional view (a); and with parallel Planes I and III (b).

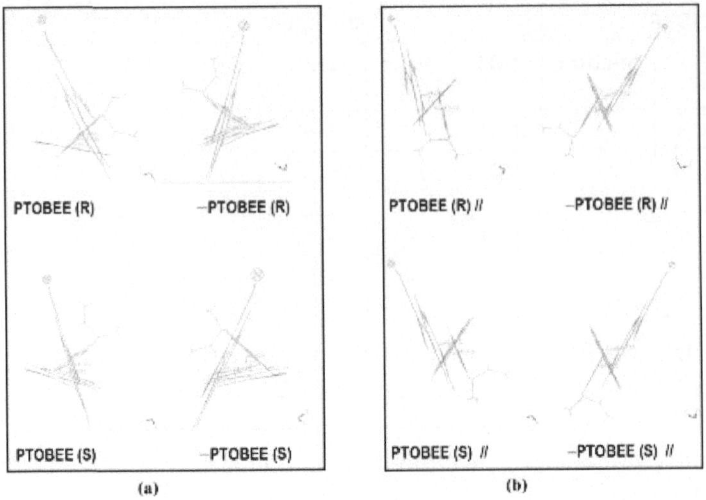

Figure 13. Conformers of monomeric PTOBEE with both R and S chirality, with triangular sectional view (a); and with Planes I and III parallel (b).

Table 3. Torsional angles (°) for the four possible helical conformations of PNOBDME and PTOBDME monomers: Rgg, Rgt, Sgg, Sgt, after energy minimization.

PNOBDME		Measogen				SPACER			E (KCAL/ MOL)
Conformer	CNI	NIC_B	N2C	C_BN2	CO	PHI (φ)	TAU	C*-CH	
R_{gg}	-178.5	47.81	175.5	48.56	-180	-75.6	-52.0	161.66	-63.53
R_{gt}	180	50.09	179	38.5	-180	-76	-52	161.2	-64.5
S_{gg}	178.07	-47.6	-175.4	-48.07	180	75.04	51.14	-162.3	-63.53
S_{gt}	-177	-34.5	-177	-40.4	173.6	92.43	56.34	-144.81	-65.53

PNOBDME		Measogen				SPACER			E (KCAL/ MOL)
Conformer	CNI	NIC_B	N2C	C_BN2	CO	PHI (φ)	TAU	C*-CH	
R_{gg}	-170.5	70.4	175.3	48	162.6	-61.83	-59.5	179	-31.6
R_{gt}	-176.1	72.9	177.6	63.8	179.6	-80.2	-53.35	156.7	-21.94
S_{gg}	177.7	-70	177.8	-69.3	-179.6	80.44	55.3	-156.47	-21.89
S_{gt}	-178	70.4	-178.2	70.9	-174.6	81	-176.3	-156.4	-20.3

2.2 Conformational analysis of PNOBDME and PTOBDME

Table 3 shows the independent torsional values of the four possible staggered conformers of PNOBDME monomer: Rgg, Rgt, Sgg, Sgt, compared to those of PTOBDME, after energy was minimized by geometry optimization.

All the torsion angles values of PNOBDME Rgg conformer are quite similar to those of conformer Sgg but with opposite signs. The same is observed between Rgt and Sgt values.

PNOBDME Rgg and Rgt conformers exhibit the same torsions sign and even similar values of the head torsions within the spacer: CO, Phi, Tau and C*-CH. The same similarity is observed between Sgg and Sgt but with opposite signs.

A systematic Conformational Analysis (CA) applied to the PTOBDME monomer for the two torsions (CO1, CBO1) in one terephthalate ester bridge, had given four equi-energetic minima, observed in the (CO, CBO) map, Figure 14, at points: A (163°; 44°), B (163°; 44°), C (163°; 141°), D (163°; 141°) [25], very close values to those found for PTOBEE in Table 2. Other secondary minima were found at E(42°; 36°) and F(41°; 144°) with slightly higher energy. The same results were obtained in the CA of the second terephthalate ester group with the two torsions (CO2, CBO2).

The systematic CA for the four torsions defined for the spacer (head) in PTOBDME monomer, CO (i.e., Ccarbonyl Oester), Phi, Tau, and C* CH, had given 57 sterically allowed conformations after removal of duplicates, with an energy threshold of 10 kcal/mol. The three optimized conformers with the lowest energy, E = 74,2 kcal/mol, resulted in a unique solution at (CO, Phi, Tau, C* CH) = (177°; 77°; 64°; 179,75°) [25]. These values are quite similar to those found for the four staggered diastereomeric conformers of PNOBDME in Table 3.

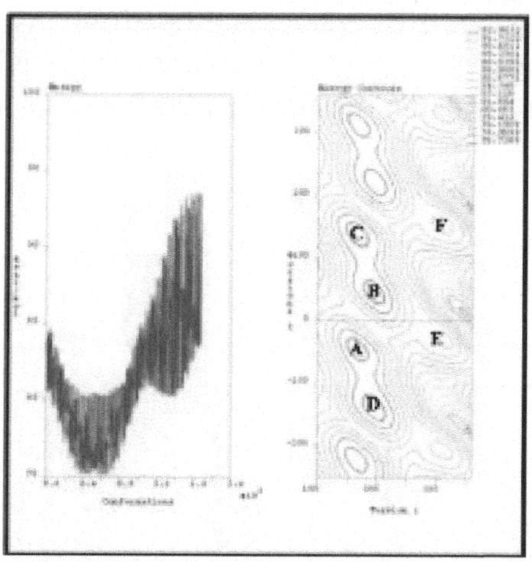

Figure 14. Systematic conformational search map for torsions: 1ζ = CO1, 2ζ = CBO1 in PTOBDME, after [25].

An ordinary molecular dynamics simulation had also been carried out with the PTOBDME monomer, including all torsions, imposing temperatures between 1600 and 500 K, followed by Quenching or Annealing. The lowest energy minima found included the head torsions (-175°; 77°; -57°; -174°) [25]. Similar to those found for PTOBEE (172,6°; -72,88°; -51,9°; 163,8°) in Table 2.

3. Polymer chains

Homopolymerization of the monomers with minimized energy was then simulated with Materials Studio® [30] by imposing Head-to-Tail orientation, torsion angle between monomers fixed to 180° and isotacticity to the polymer chain.

Helical rigid/semirigid chains models were obtained in all LC polymers with the chiral groups located in the spacer along the main backbone chain, and with the flexible lateral chains located either far or near the helix, one of the major types of macromolecules forming the cholesteric mesophase [32], as schematized in Figure 15. The total energy of the polymer models was also minimized and their geometry optimized with Forcite module of Materials Studio® [30].

The lowest energy helical molecular model of Poly(PTOBEE-S //)25 with pentagonal cross-section is shown in Figure 16 as an example.

Molecular models of the four helical conformers of polyesteramide (PNOBDME)10, oriented parallel to the z-axis, are shown in Figure 17. Sectional-cross view, along the z-axis, of Poly(PNOBDME Rgg)10; Poly(PNOBDME Sgg)10; Poly(PNOBDME Rgt)10 and Poly(PNOBDME Sgt)10.

The main structural difference among the PNOBDME conformers is the different positions of the lateral chains. In conformers Rgg and Sgg the side chains are allocated among the benzoyl groups, with the result of its restricted mobility, in agreement with the positive NOE values obtained for precursor PTOBDME [16]. While in Rgt and Sgt conformers the lateral chains are placed outside the mesogenic benzoyl groups, exhibiting higher mobility, with a negative NOE.

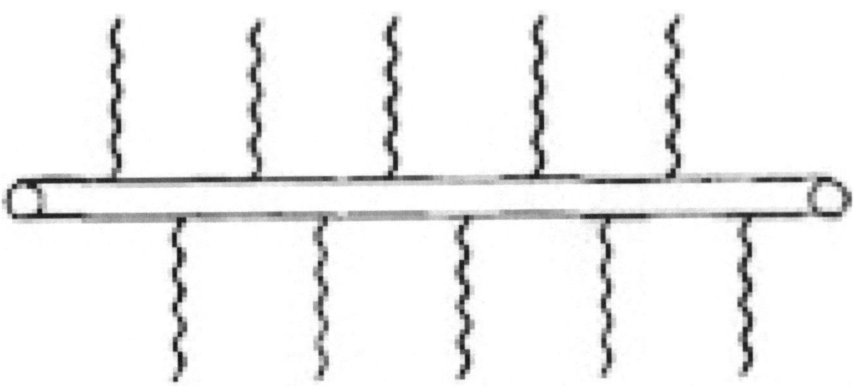

Figure 15. Schematic representation of the cholesteric LC polymer type, with a rigid or semirigid helical chain with flexible branches.

Figure 16. Sectional cross-view of Poly(PTOBEE-S //)25.

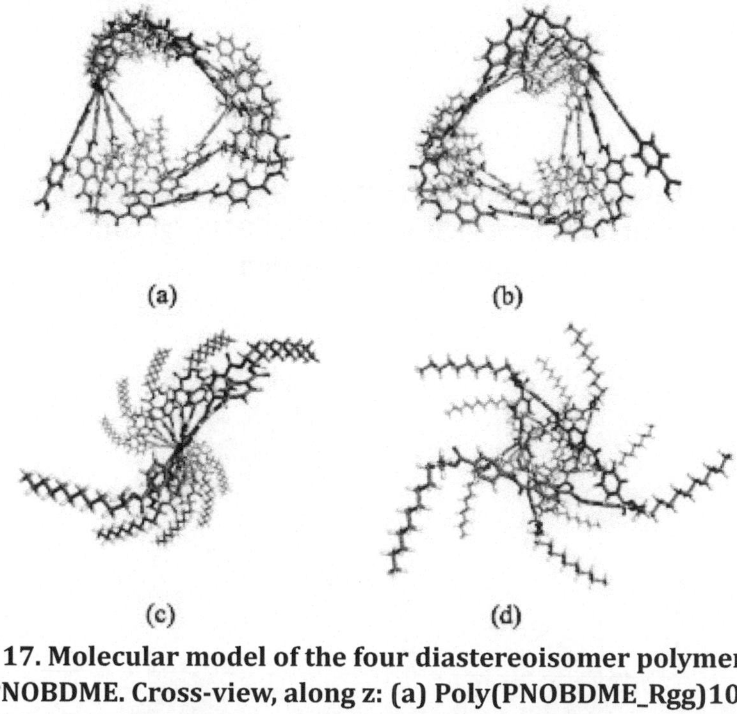

Figure 17. Molecular model of the four diastereoisomer polymer chains of PNOBDME. Cross-view, along z: (a) Poly(PNOBDME_Rgg)10; (b) Poly(PNOBDME_Sgg)10; (c) Poly(PNOBDME_Rgt)10; (d) Poly(PNOBDME_Sgt)10.

The same effect has been observed for polyester PTOBDME Sgg and Sgt conformers. Their lateral view and cross-section can be seen in Figure 18.

The four helical conformers models of polyesteramide (PNOBDME)100: Poly(PNOBDME_Rgg)100; Poly(PNOBDME_Sgg)100; Poly(PNOBDME_Rgt)100 and

Poly(PNOBDME_Sgt)100, are simulated in Figure 19, along the z-axis (sectional cross view).

Their lateral view is reproduced in Figure 20. A third order of chirality corresponding to the twisting of the rigid/semirigid cholesteric LC polymer chain can be observed.

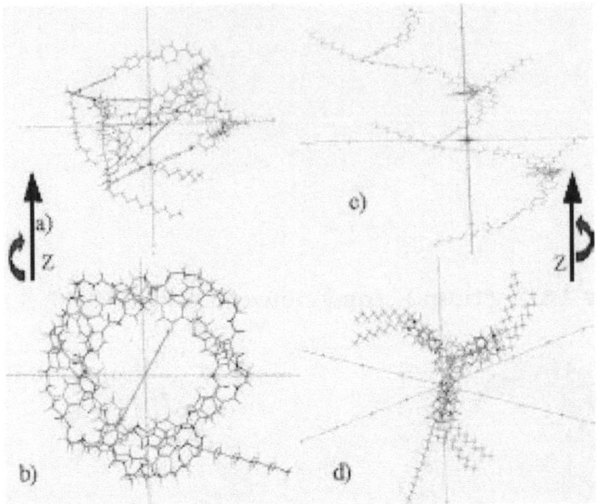

Figure 18. (a) Lateral view of PolyPTOBDME_Sgg conformer along Z axis and (b) cross section; (c) lateral view of PolyPTOBDME Sgt conformer along Z and; (d) cross section.

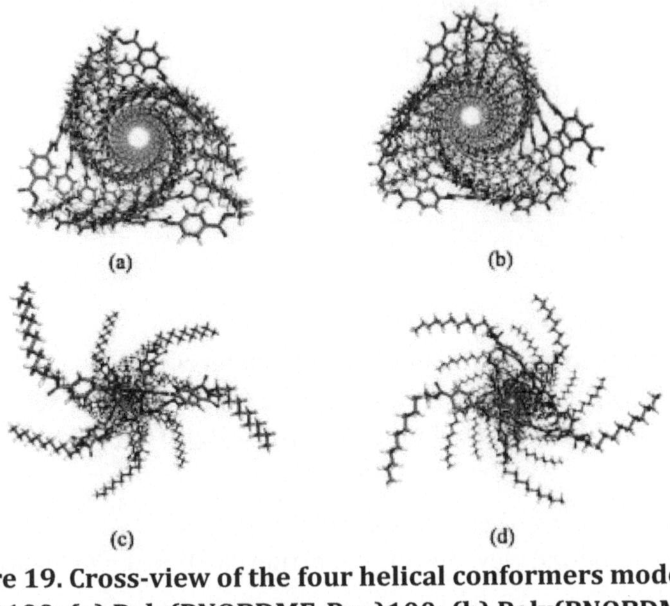

Figure 19. Cross-view of the four helical conformers models of (PNOBDME)100: (a) Poly(PNOBDME_Rgg)100; (b) Poly(PNOBDME_Sgg)100; (c) Poly(PNOBDME_Rgt)100 and (d) Poly(PNOBDME_Sgt)100.

4. Crystalline structure and morphology modelization of precursor polyesters PTOBEE and PTOBDME

4.1 Crystal structure simulation of precursor polyester PTOBEE

Crystal packing of helical molecular chain model of (PTOBEE-R)8 (as defined in Table 2) has been simulated, with Materials Studio® [30], in a triclinic primitive P1 unit cell, Figure 21b, with parameters: a = 5,6 Å, b = 4,65 Å, c = 177,2 Å (equal to the pitch length of each simulated polymer helix), α = 90°, β = 120°, and γ = 88°.

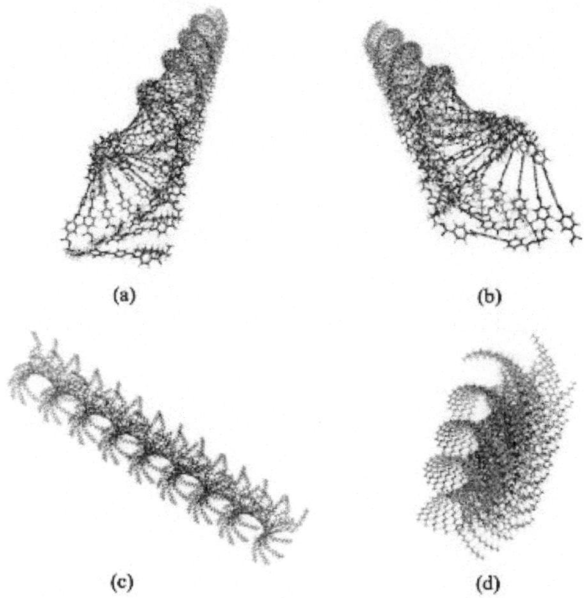

Figure 20. Lateral view of the polymer models of the four helical conformers of (PNOBDME)100. (a) Poly(PNOBDME_Rgg)100; (b) Poly(PNOBDME_Sgg)100; (c) Poly(PNOBDME_Rgt)100 and (d) Poly(PNOBDME_Sgt)100.

Polymer molecular models were oriented with their main chain parallel to the z-axis. Only one cell has been drawn. Its content is repeated in the x and y directions with a and b translational vectors to build a 3D polymer crystalline model.

The calculated X-ray diffraction pattern of PTOBEE from the crystal model is shown in Figure 21d compared to the experimental Synchrotron X-Ray diffraction pattern of PTOBEE under dynamical conditions in Figure 21c, where two reflections always present throughout the experiment at 2θ 16° and 24° were associated with the mesophase state and another two at 19° and 28° with the crystal formation. The relative intensity of peaks assigned to crystal presence is lost, after the temperature of the transition crystal to liquid-crystal is reached, at about 150°C, during the heating process; and they almost do not recover after the cooling process as compared with isothermal crystallization. The molecular model of the polymer (PTOBEE-R)66 (in Table 2) is displayed in Figure 21e, exhibiting square cross-section, and the twisting of the helical chain at higher level. Experimental microphotographs of PTOBEE single crystals are shown in Figure 21f and g.

4.2 Crystal structure simulation of precursor polyester PTOBDME

The helical molecular model of Poly(PTOBDME)n could be packed in one triclinic primitive P1 unit cell with parameters: a = 5,5 Å, b = 4,7 Å, c (helical pitch length depending on the model) =222,1 Å for Poly(PTOBDME_Rgg)40, c = 169,8 Å for Poly(PTOBDME_Rgt)15, c = 171,5 Å for Poly(PTOBDME_Sgg)15, c = 277,5 Å for Poly(PTOBDME_Sgt)40, α = 92°, β = 112°, and γ = 90°.

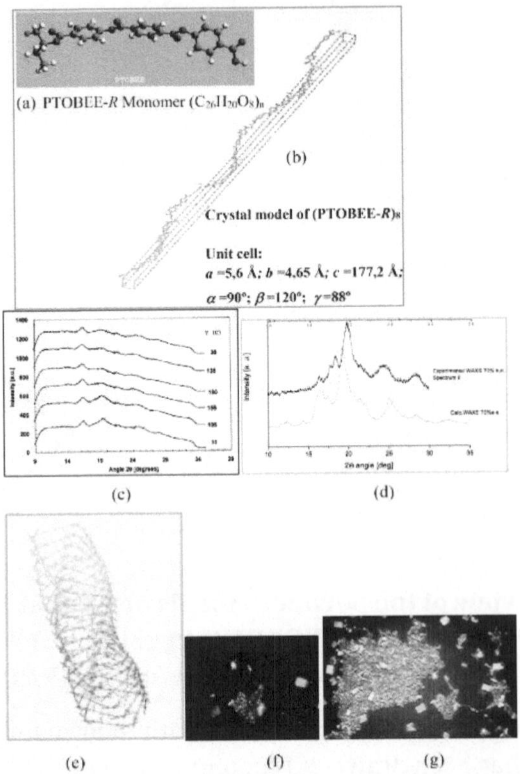

Figure 21. Structure of precursor polyester (PTOBEE)8: (a) monomer model (C26H20O8); (b) simulated crystal structure; (c) experimental synchrotron X-Ray diffraction pattern of PTOBEE under dynamical conditions: Heating range between 30°C and 180°C at 10°C/min,5 min, kept isothermally at 180°C, and cooling range from 180–30°C at 10°C/min; (d) Calculated X-ray diffraction pattern of PTOBEE from the crystal model, refined to match the experimental; (e) molecular model of (PTOBEE-R)66 (in Table 2) with square cross-section and with helix writhing at a higher level; (f) and (g) experimental crystal photographs of PTOBEE.

The morphology of the polymer crystal model was further simulated also with Materials Studio® [30]. The simulated crystal morphology appears in Figure 22b with a rombohedrical pink shape. It is in good agreement with the rhombohedral crystals observed by optical microscopy, dispersed within the mesophase matrix, Figure 22d.

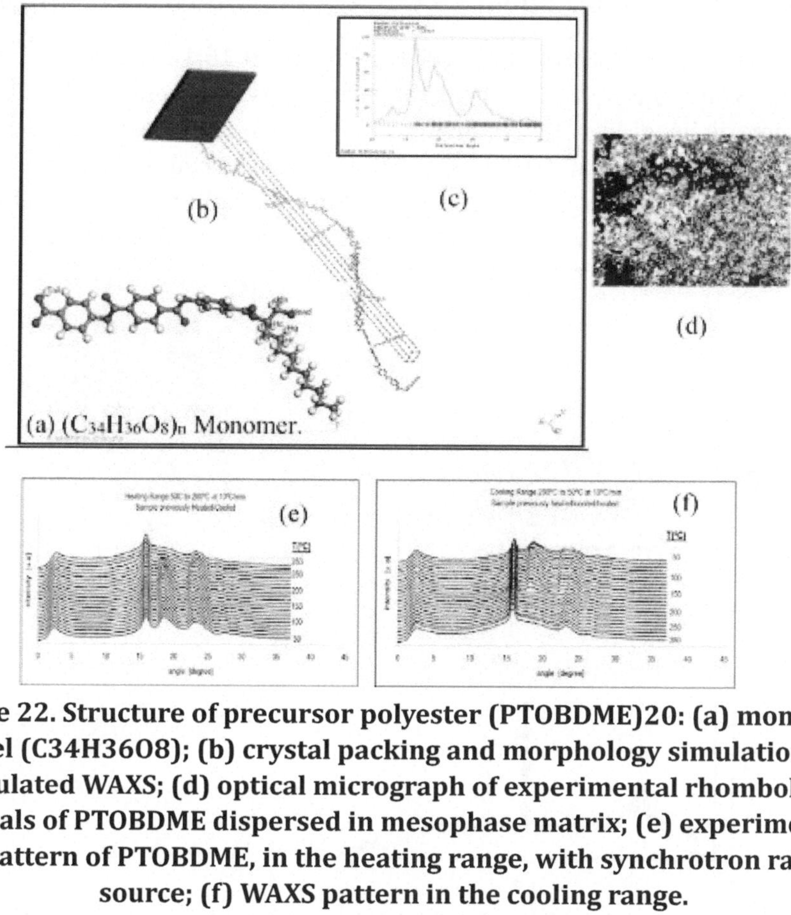

Figure 22. Structure of precursor polyester (PTOBDME)20: (a) monomer model (C34H36O8); (b) crystal packing and morphology simulation; (c) recalculated WAXS; (d) optical micrograph of experimental rhombohedral crystals of PTOBDME dispersed in mesophase matrix; (e) experimental WAXS pattern of PTOBDME, in the heating range, with synchrotron radiation source; (f) WAXS pattern in the cooling range.

Experimental XRD patterns of PTOBDME registered during heating and cooling ranges, with synchrotron radiation source, are given in Figure 22e and f respectively. Two XRD peaks, at 2θ 16° (d100 = 5,5 Å) and 24° (3,71 Å), always present in the entire temperature range, were assigned to the cholesteric mesophase. The third one at 2θ 19° (d010 = 4,7 Å) and a small one at 2θ 22,5° (3,95 Å), disappearing at 240°C during the heating range and appearing again at about 150°C while cooling, were interpreted as due to the presence of a three-dimensional (3D) order, in agreement with the crystalline stability temperature, DSC curves. This means that a cholesteric mesophase is always present and coexists with the crystalline phase.

5.Crystalline structure and morphology. Modelization of polyesteramide PNOBDME

5.1 Experimental simultaneous SAXS/WAXS patterns of PNOBDME

The powder SR diffraction data were registered at 16.1 beamline at Daresbury Laboratory, Warrington, U.K., during the heating range from 30–200°C, with a monochromatic beam (λ = 1,4 Å), Figure 23. Both SAXS and WAXS detectors were lineal. Wet collagen (rat tail tendon, d = 676,08 Å) was used to calibrate the q-axis (Å-1) of the SAXS detector. The WAXS detector was calibrated with HDPE, with the scattering angle defined by 2θ. The experimental patterns were corrected for background scattering, with the help of ATSAS 3.0 [33].

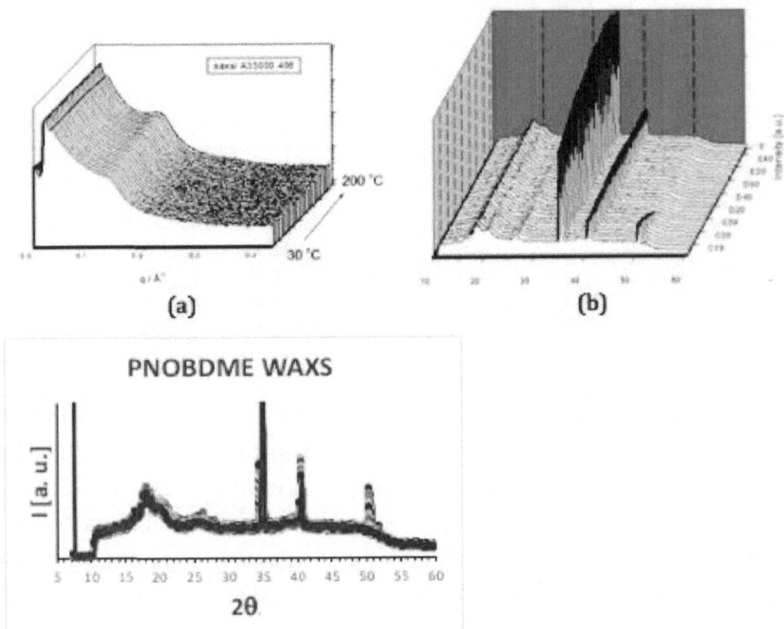

(c)

Figure 23. Experimental simultaneous SAXS (a)/WAXS (b) patterns of PNOBDME with synchrotron radiation source; (c) projection of all superimposed WAXS patterns.

The SAXS spectra in Figure 23a show two sharp order reflections at q value of 0,018 Å-1 (55,5 Å) and 0,029 Å-1 (34,48 Å).

Four WAXS peaks, Figure 23b at 2θ 18,39° (d = 4,38 Å); 18,83° (d = 4,28 Å); 35,17° (d = 2,31 Å) and 40,75° (d = 2,01 Å) are systematically observed in the entire temperature range, and were assigned to the cholesteric mesophase.

While peaks at 2θ 26,39° (d = 3,07 Å); 50,71° (d = 1,63 Å) disappearing at about 90°C during the heating range are attributed to crystal 3D phase.

5.2 Crystal structure simulation of polyesteramide PNOBDME

Crystal packing for each of the four helical diastereoisomer polymer chains of Poly(PNOBDME)10 has been simulated, with Materials Studio® [30], in triclinic primitive P1 unit cells, with their main chains oriented parallel to the z-axis, Figure 24. Only one unit cell has been drawn. Its content is repeated in the x and y directions with a and b translational vectors to build a 3D polymer crystalline model.

Crystal models have also been simulated for the four PNOBDME diastereoisomers with different degrees of polymerization, n = 10, 30, 60, 100, and optimized by energy minimization with the Forcite module [30], as shown in Table 4.

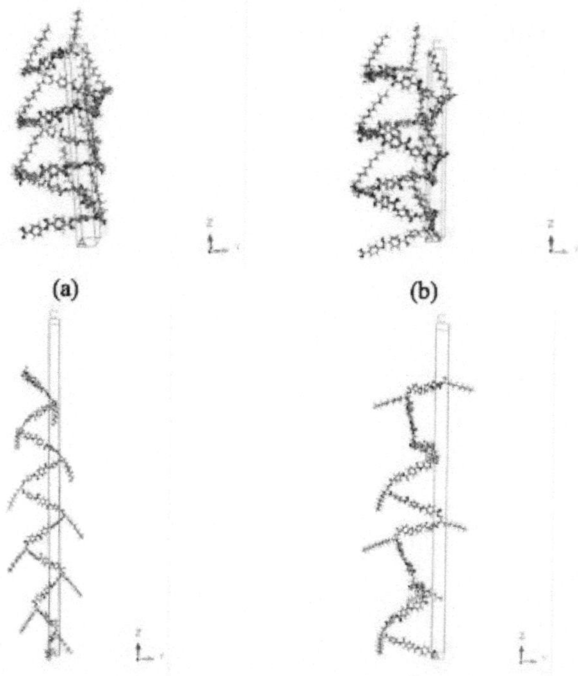

Figure 24. Crystal model details of the four helical conformers of PNOBDME: (a) Poly(PNOBDME_Rgg)10; (b) Poly(PNOBDME_Sgg)10; (c) Poly(PNOBDME_Rgt)10; (d) Poly(PNOBDME_Sgt)10.

5.3 Structure refinement

From the different simulated crystal models, their theoretical powder diffraction patterns can be calculated, and the polymer structure refined until the simulated spectra match the experimental diffraction pattern of PNOBDME, in Figure 23b. Changes in positional parameters cause changes in structure-factor magnitudes and therefore in relative peak intensities, whereas atomic displacement (thermal) parameters have the effect of emphasizing the high-angle region (smaller thermal parameters) or de-emphasizing it (larger thermal parameters). It is usually advisable to start the refinement of structural parameters with the positions of the heavier atoms and then to try those of the lighter atoms. If the latter refinement converges, all atomic positions in the model can then be refined simultaneously. At this point, the refinement of the somewhat trickier parameters can be attempted [34, 35].

5.3.1 Pawley refinement

Crystal_Poly(PNOBDME_Rgg)10 model, with a minimized total energy = _244 kcal/mol, has been first refined by successive Pawley cycles, by optimizing the Lattice Parameters (a, b, c, α, β, γ); Pattern Parameters (Pseudo-Voigt profile function): Peaks FWHM (U, V, W), Profile parameters (NA, NB), Line shift (Instrument Geometry Bragg–Brentano): Zero point value, Shift #1, Shift #2; Sample parameters: Crystallite size broadening (A, B, C).

Table 4. Pawley refinement optimized results, minimized E, refined lattice cell parameters, agreement indices (R values).

Crystal diastereoisomer	Minimized_E (kcal/mol)	a (Å)	b (Å)	c (Å)	α (°)	β (°)	γ (°)	R_{wp}, R_p, $R_{wp(w/o\ bc)}$, CMACS Pawley
Poly(PNOBDME_R*gg*)$_{10}$	—224,96	5,919	5,04	55,29	90,57	109,00	89,50	0,44%;0,09%;0,23%;0,00%
Poly(PNOBDME_R*gg*)$_{30}$	—733,6	5,899	5,039	168,0	90,49	109,03	89,39	0,22%;0,04%;0,09%; 0,00%
Poly(PNOBDME_R*gg*)$_{60}$	—1.459,5	6,00	5,00	330,0	90,6	109,0	89,0	0,00%;0,00%;0,00%; 0,00%
Poly(PNOBDME_R*gg*)$_{100}$	—2.397,7	6,00	5,0	553	90	109	90	0,00%;0,00%;0,00%; 0,00%
Poly(PNOBDME_R*gt*)$_{10}$	—217,6	5,499	4,699	150,0	91,998	111,99	90,00	1,16%; 0,25%;0,63%;0,14%
Poly(PNOBDME_R*gt*)$_{30}$	—747,04	5,49	4,7	450	92	112	90	0,09%;0,10%;0,04%;
Poly(PNOBDME_R*gt*)$_{60}$	—1.444,5	5,5	4,7	900	92	112	90	0,00%;0,00%;0,00%; 0,00%
Poly(PNOBDME_R*gt*)$_{100}$	—2.377	5,5	4,7	1500	92	112	90	0,00%;0,00%;0,00%; 0,00%
Poly(PNOBDME_S*gg*)$_{10}$	—209,86	5,91	4,85	63,14	92,46	109,73	90,78	1,21%;0,27%;0,84%;0,28%
Poly(PNOBDME_S*gg*)$_{30}$	—617,24	5,91	4,85	189,4	92,46	109,73	90,78	0,84%;0,15%;0,42%;0,00%
Poly(PNOBDME_S*gg*)$_{60}$	—1.212	5,91	4,85	378,8	92,46	109,73	90,78	0,00%;0,00%;0,00%; 0,00%
Poly(PNOBDME_S*gg*)$_{100}$	—2.010,03	5,91	4,85	631,4	92,46	109,73	90,78	0,00%;0,00%;0,00%; 0,00%
Poly(PNOBDME_S*gt*)$_{10}$	—185,93	5,5	4,70	125,0	92,00	112,00	89,99	1,19%;0,26%;0,53%;0,25%
Poly(PNOBDME_S*gt*)$_{30}$	—531,39	5,5	4,69	375	92	112,00	89,99	0,05%;0,03%;0,05%;0,00%
Poly(PNOBDME_S*gt*)$_{60}$	—1.050,72	5,5	4,7	750	92	112	90	0,00%;0,00%;0,00%; 0,00%
Poly(PNOBDME_S*gt*)$_{100}$	—1.742	5,5	4,70	1.250	92,00	112,00	90	0,00%;0,00%;0,00%; 0,00%

The refined crystal model gave a lattice cell: a = 5,91996 (± 0,00029) Å, b = 5,04000 (± 0,00025) Å, c = 55,29001 (± 0,00284) Å, α = 90,57956 (± 0,00047)°, β = 109,00002 (± 0,00051)°, and γ = 89,50015 (± 0,00044)°, adjusted to a triclinic lattice cell with Space Group: P 1.

The Pawley adjustment factors for Crystal_Poly(PNOBDME_Rgg)10, between theoretical spectrum of the crystal model and the experimental diffraction pattern are: Final Rwp: 0,44%; Final Rp: 0,09%; Final Rwp (without background): 0,23%;

Final CMACS: 0,00%. Figure 25 displays the simulated Intensity plot vs 2θ for the crystal, compared to the experimental powder diffraction pattern, and the difference between them.

In Table 4 the optimized results are given for the crystals built from the four helical conformers: Crystal_Poly(PNOBDME_Rgg)n, Crystal_Poly(PNOBDME_Rgt) n, Crystal_Poly(PNOBDME_Sgg)n and Crystal_Poly(PNOBDME_Sgt)n, at different degrees of polymerization, n = 10, 30, 60, 100, with minimized energy, after Pawley refinement cycles of the lattice cell parameters and agreement indices values: Rwp, Rp, Rwp (without background and Final CMACS. Slightly higher a and b cell parameters are observed for Rgg and Sgg conformers with broader chain diameters.

5.3.2 Rietveld refinement

Starting from the best Pawley refined model, the first cycle of Rietveld is accomplished by optimizing the Cell Lattice Parameters (a, b, c, α, β, γ) into a triclinic unit cell with Space Group P1; Pattern Parameters (Pseudo-Voigt peak Profile Function): FWHM (U, V, W), profile parameters (NA, NB), Line Shift parameters (instru- ment Geometry Bragg–Brentano): Zero point, Shift #1, Shift #2, Asymmetry (Rietveld correction) P parameter, Background coefficients (Polynomial order = 20); Sample Parameters: Crystallite Size broadening (A, B, C), Preferred Orientation (Function Rietveld-Toraya): (a*, b*, c*, G1, G2), Atomic Temperature Factors applied; Structure parameters, 30 torsions being refined (hydrogen atoms used); Atoms Temperature Factors set to anisotropic, first for only the N and O atoms and after for all the atoms, Occupancies are also refined.

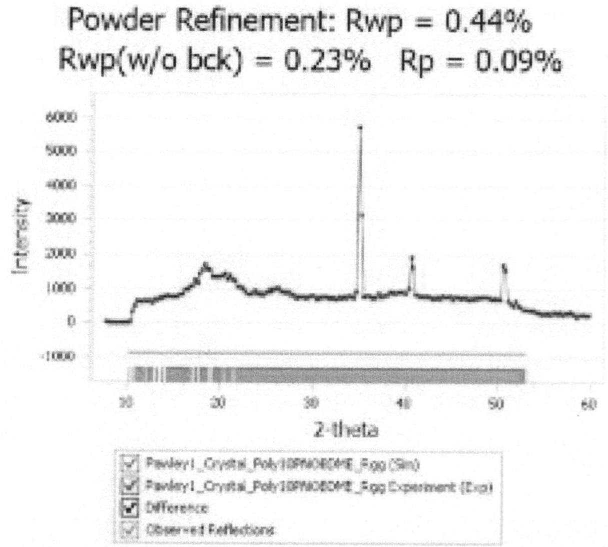

Figure 25. Pawley refinement of synchrotron powder diffraction pattern of Crystal_Poly(PNOBDME_Rgg)10. Observed intensities-magenta-line, calculated intensities-blue points. Green ticks are reflection positions. The difference (observed-calculated) curve-black line.

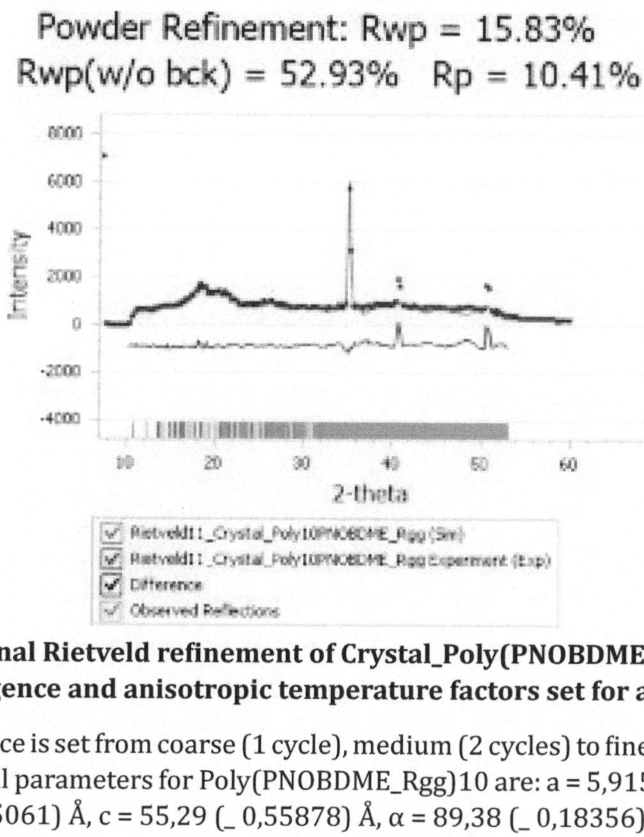

Figure 26. Final Rietveld refinement of Crystal_Poly(PNOBDME_Rgg)10 with fine-convergence and anisotropic temperature factors set for all the atoms.

Convergence is set from coarse (1 cycle), medium (2 cycles) to fine (4 cycles). The refined unit cell parameters for Poly(PNOBDME_Rgg)10 are: a = 5,915 (_ 0,05811) Å, b = 5,03 (_ 0,05061) Å, c = 55,29 (_ 0,55878) Å, α = 89,38 (_ 0,18356)°, β = 109,01 (_

0,26747)°, and γ = 88,91 (_ 0,23494)°. The adjustment factors: Final Rwp: 15,83% Final Rp: 10,41% Final Rwp (without background): 52,92% Final CMACS: 6,62%, in Figure 26..

5.4 Powder quantitative phase analysis

The Powder Quantitative Phase Analysis module of Materials Studio® [30] has been used to determine the relative amounts of different phases in a mixture from a powder diffraction pattern of the mixture.

An approximation has been made considering ideal mixtures with the four helical conformers crystal structures of Crystal_Poly(PNOBDME)n as pure component phases, for each degree of polymerization, n = 10, 30, 60, 100, their relative amounts being determined from Experimental PNOBDME Powder diffraction pattern, Figure 23c.

The Refined weights for Crystal_Poly(PNOBDME)10 pure phases: Rgg, Sgg, Rgt and Sgt in, Figure 27, gave Crystal_Poly(PNOBDME)10_Sgg as preferable major phase (90,79%) and Crystal_Poly(PNOBDME)10_Rgg in lower proportion (9,199%).

The Refined weights for Crystal_Poly(PNOBDME)30 pure phases: Rgg, Sgg, Rgt and Sgt, in Figure 28, gives Crystal_Poly(PNOBDME)30_Rgg as major phase (86,45%), with Crystal_Poly(PNOBDME)30_Rgt (13,54%).

Powder QPA Reflex QPA Job\Powder QPA *

	A	B	C	D	E	F
	Phase ID	Cell formula	Structure	Integrated intensity	Intensity %	Weight %
1	Crystal_Poly10PNOBDME_Rgg	C340 H382 N20 O60	Pawley1_Crystal_Poly10PNOBDME_Rgg	12,20945931	0,34633962	9,19970901
2	Crystal_Poly10PNOBDME_Sgg_	C340 H382 N20 O60	Crystal_Poly10PNOBDME_Sgg_GeomOpt_	95,46238172	2,70793356	90,79839978
3	Crystal_Poly10PNOBDME_Rgt_	C340 H372 N20 O60	Crystal_Poly10PNOBDME_Rgt_GeomOpt_	0,00132172	3,749247e-005	0,00189121
4	Crystal_Poly10PNOBDME_Sgt_	C340 H382 N20 O60	Crystal_Poly10PNOBDME_Sgt_GeomOpt_	2,252517e-010	6,389601e-012	1,390084e-010

Figure 27. Refined weights on ideal mixture of Crystals_Poly(PNOBDME)10 pure phases: Rgg, Sgg, Rgt and Sgt.

QPA_Crystal_Poly30PNOBDME Reflex QPA Job\QPA_Crystal_Poly30PNOBDME *

	A	B	C	D	E	F
	Phase ID	Cell formula	Structure	Integrated intensity	Intensity %	Weight %
1	Crystal_Poly30PNOBDME_Rgt	C1020 H1142 N60 O180	Crystal_Poly30PNOBDME_Rgt	57,11112453	1,47345357	13,54584254
2	Crystal_Poly30PNOBDME_Sgg_	C1020 H1142 N60 O180	Crystal_Poly30PNOBDME_Sgg_	7,481918e-009	1,930317e-010	1,701925e-009
3	Crystal2_Poly30PNOBDME_Sgg	C1020 H1142 N60 O180	Crystal2_Poly30PNOBDME_Sg	7,328505e-009	1,890737e-010	1,745568e-009
4	Crystal_Poly30PNOBDME_Rgg_	C1020 H1142 N60 O180	Crystal_Poly30PNOBDME_Rgg_	391,01486639	10,08809154	86,45415746

Figure 28. Refined weights of Crystal_Poly(PNOBDME)30 pure phases: Rgg, Sgg, Rgt and Sgt.

QPA_Crystal_Poly60PNOBDME Reflex QPA Job\QPA_Crystal_Poly60PNOBDME *

	A	B	C	D	E	F
	Phase ID	Cell formula	Structure	Integrated intensity	Intensity %	Weight %
1	Crystal_Poly60PNOBDME_Rgt	C2040 H2282 N120 O360	Crystal_Poly60PNOBDME_Rgt	5,073964e-009	1,318969e-010	1,486595e-009
2	Crystal_Poly60PNOBDME_Sgg_	C2040 H2282 N120 O360	Crystal_Poly60PNOBDME_Sgg_	2,538874e-011	6,595685e-013	7,449604e-012
3	Crystal_Poly60PNOBDME_Sgt	C2040 H2282 N120 O360	Crystal_Poly60PNOBDME_Sgt	422,04715800	10,96772783	100,00000000
4	Crystal_Poly60PNOBDME_Rgg_	C2040 H2282 N120 O360	Crystal_Poly60PNOBDME_Rgg_	2,853625e-009	7,415706e-011	8,883851e-010

Figure 29. Refined weights of Crystal_Poly(PNOBDME)60 pure phases: Rgg, Sgg, Rgt and Sgt.

QPA_Crystal_Poly100PNOBDME Reflex QPA Job\QPA_Crystal_Poly100PNOBDME *

	A	B	C	D	E	F
	Phase ID	Cell formula	Structure	Integrated intensity	Intensity %	Weight %
1	Crystal_Poly100PNOBDME_Sgg_	C3400 H3802 N200 O600	Crystal_Poly100PNOBDME_Sgg	2,554348e-013	7,282688e-015	5,957682e-013
2	Crystal_Poly100PNOBDME_Sgt	C3400 H3802 N200 O600	Crystal_Poly100PNOBDME_Sgt	63,37894270	1,80698868	99,99999912
3	CrystalPoly100PNOBDME_Rgt_C	C3400 H3802 N200 O600	CrystalPoly100PNOBDME_Rgt_(	3,366750e-009	9,598896e-011	7,918773e-009
4	Crystal_Poly100PNOBDME_Rgg_	C3400 H3802 N200 O600	Crystal_Poly100PNOBDME_Rgg	3,723894e-007	1,061715e-008	8,947581e-007

Figure 30. Refined weights of Crystal_Poly(PNOBDME)100 pure phases. Rgg, Sgg, Rgt and Sgt.

The Refined weights for Crystal_Poly(PNOBDME)60 four pure phases, in Figure 29, gives Crystal_Poly(PNOBDME)60_Sgt as major component (100%).

The Refined weights for Crystal_Poly(PNOBDME)100 pure phases, in Figure 30, gives Crystal_Poly(PNOBDME)100_Sgt as major component (100%).

These results would redound to the previously postulated kinetic resolution of a preferable helical diastereomer, with respect to the four possible, the R/S ratio of asymmetric carbon atoms remaining 50:50, depending on the synthetic conditions and the degree of polymerization.

The results presented here are only indicational; more refinement cycles would be needed to improve accuracy.

6. Conclusions

Conformational and Structure Analysis has been performed on polyesteramides PNOBEE and PNOBDME compared to those of polyesters precursor.

Since the mesogen of PNOBDME and PNOBEE consists of three planar rigid groups: one terephthalamide and two benzoates, connected by amide and ester groups, a simplified model could be reduced to three articulated planes I, II and III, in the molecular scheme.

Three orders of chiralities are observed: One due to the asymmetric carbon atoms, a second chirality due to the two successive rotations of the benzene groups within the monomer, along the main chain, which imply the formation of helical molecules with either the three benzene groups forming a triangular cross-section (Δ) or with Planes I and III becoming parallel (//), for both R and S chirality.

Moreover, a third chirality corresponds to the twisting of the rigid/semirigid cholesteric LC polymer chain LC polymer.

Four helical conformer models are proposed for each cholesteric polyesteramide: Rgg, Rgt, Sgg, Sgt, experimentally observed by NMR.

Polymerization of the monomeric conformers, with minima energies, have been simulated and used to reproduce the crystalline fraction observed by x-ray diffraction.

Crystal models of the four PNOBDME diastereoisomers have been simulated with different degrees of polymerization, n = 10, 30, 60, 100, and optimized by energy minimization with the Forcite module.

Crystal packing was simulated in triclinic primitive P1cells, with molecular chains oriented parallel to the z-axis (with c lattice parameter attained from simulated "polymerization" and parameters a, b, α, β and γ, obtained by Pawley refinement from the known structures of precursor polyesters.

Successive Pawley refinement cycles optimized the lattice cell parameters for all the simulated polymers, obtaining excellent adjustment factors for Crystal_Poly(PNOBDME_Rgg)10, between theoretical spectrum of the crystal model and the experimental diffraction pattern are: Final Rwp: 0,44%; Final Rp: 0,09%; Final Rwp (without background): 0,23%; Final CMACS: 0,00%.

Sequential Rietveld cycles refined the structures with fine-convergence, by changing positional parameters 30 torsion angles, causing changes in structure-factor magnitudes, relative peak intensities, and atomic displacement (thermal parameters) set first isotropic for all the atoms and after set to aniso- tropic for only the N and O atoms and then for the rest. Occupancies are also refined. Final Rwp: 15.83% Final Rp: 10.41% Final Rwp (without background): 52.92% Final CMACS: 6.62.

Optimized results are given for the crystals built from the four helical conformers: Crystal_Poly(PNOBDME_Rgg)n, Crystal_Poly(PNOBDME_Rgt)n, Crystal_Poly(PNOBDME_Sgg)n and Crystal_Poly(PNOBDME_Sgt)n, at different degrees of polymerization, n = 10, 30, 60, 100, with minimized energy, after Pawley refinement cycles of the lattice cell parameters and agreement indices values: Rwp, Rp, Rwp.

Crystal structure models are proposed whose simulated x-ray diffraction patterns fit the experimental WAXS, after successive Pawley and Rietveld refinement cycles.

A Powder Quantitative Phase Analysis approximation has been made considering ideal mixtures with the four helical conformers crystal structures of Crystal_Poly(PNOBDME)n as pure component phases, for each degree of polymerization, n = 10, 30, 60, 100, their relative amounts being determined from Experimental PNOBDME Powder diffraction pattern.

The Refined weights for Crystal_Poly(PNOBDME)10 pure phases: Rgg, Sgg, Rgt and Sgt, gave Crystal_Poly(PNOBDME)10_Sgg as preferable major phase (90,79%) and Crystal_Poly(PNOBDME)10_Rgg in lower proportion (9,199%).

The Refined weights for Crystal_Poly(PNOBDME)30 pure phases: Rgg, Sgg, Rgt and Sgt, gives Crystal_Poly(PNOBDME)30_Rgg as major phase (86,45%), with Crystal_Poly(PNOBDME)30_Rgt (13,54%).

The Refined weights for Crystal_Poly(PNOBDME)60 and Crystal_Poly (PNOBDME)100, four pure phases, give Crystal_Poly(PNOBDME)60_Sgt as major component (100%).

These results confirm the previously postulated kinetic resolution of a preferable helical diastereomer, with respect to the four possible, depending on the synthetic conditions and the degree of polymerization, while the R/S ratio of asymmetric carbon atoms remains 50:50.

References

1 McMillan F M. The Chain Straighteners. Fruitful Innovation: The Discovery of linear and Stereoregular Synthetic Polymers: The Macmillan Press LTD 1979, London.

2 Carothers W H. Studies on polymerization and ring formation. I. An introduction to the general theory of condensation polymers. J. Am. Chem. Soc. 1929; 51, 8: 2548-2559.

3 Pino P, Ciardelli F, Zandomeneghi M. Optical Activity in Stereoregular Synthetic Polymers Annual Review of Physical Chemistry October 1970; Vol. 21:1: 561-608. https://doi.org/10.1146/a nnurev.pc.21.100170.003021]

4 Vogl O, Corley L S, Harris W J, Jaycox G D, Zhang J. Optical activity based on macromolecular asymmetry. Makromol.Chem., Suppl. 1985; 13 :1-12.

5 Mallakpour S, Zadehnazari A. Advances in synthetic optically active condensation polymers – A review. eXPRESS Polymer Letters 2011; 5, (2): 142–181 Available online at www.expre sspolymlett.com. DOI: 10.3144/ expresspolymlett.2011.15

6 Ringsdorf H, Schlarb B, Venzmer J. Molecular Architecture and Function of Polymeric Oriented Systems: Models for the Study of Organization, Surface Recognition, and Dynamics of Biomembranes. Angew. Chem. Int. Ed. Engl. 1988; 27, 113-158.

7 International Union of Crystallography, Report of the Executive Committee for 1991, Acta Cryst. 1992; A48,922.

8 Zanchetta G. Liquid crystalline phases in oligonucleotide solutions [thesis] University of Milan 2007

9 Kornyshev A A, Leikin S, Malinin S V. Chiral electrostatic interaction and cholesteric liquid crystals of DNA. Eur. Phys. J. E 2002; 7: 83-93

10 Lai S L, Hartono D, Yang K-L. Self-assembly of cholesterol DNA at liquid crystal/ aqueous interface and its application for DNA detection. Appl. Phys. Lett. 2009; 95: 153702.

11 Pérez Méndez M. "Synthesis and Characterization of Biocompatible Polyesteramides: PNOBDME (C34H38N2O6)n and PNOBEE (C26H22N2O6)n as Cholesteric Liquid Crystals. Polymer Science 2018; 4, 2. DOI: 10.36648/2471-9935.4.2.37. https://polymerscience.imedpub.com/ synthesis-and-characterization-of- biocompatible-polyesteramides- pnobdmec34h38n2o6n-and-pnobee- c26h22n2o6n-as-cholesteric-liquid-cr. php?aid=2395

12 Pérez Méndez M. Biocompatible, Nanostructured, Chiral Polyesteramides: PNOBDME (C34H38N2O6)n and PNOBEE (C26H22N2O6)n Synthesized and Characterised as Cholesteric Liquid Crystals. International Journal of Engineering Research and Applications (IJERA) 2019; 9, 6 (Part 1): 52-66; ISSN: 2248-9622; DOI: 10.9790/9622- 0906015267.

13 Pérez Méndez M, Fayos Alcañiz J. Cholesteric Liquid Crystal Polyesteramides, Non-viral Vectors. In: Liquid Crystals and Display Technology: IntechOpen; 2020. DOI: 10.5772/intechopen.9131. Available from: https://www.intechopen.com/ online-first/cholesteric-liquid-crystal- polyesteramides-non-viral-vectors.

14 Pérez-Méndez M, Marco C. New synthesis, thermal properties and texture of cholesteric Poly[ethylethylene 4,40-(terephthaloyldioxy) dibenzoate]. Acta Polymerica 1997; 48:502-506.

15 Pérez-Méndez M, Marco Rocha C. Preparing cholesteric liquid-crystals - by adding acid dichloride and butanediol to chloro-naphthalene, heating in nitrogen, decanting into toluene, etc. Patent with n° EP1004650-A; WO9831771-A; WO9831771-A1; AU9854863-A; ES2125818-A1; ES2125818-B1; EP1004650-A1; US6165382-A; MX9906732-A1; JP2001513827-W; AU739076-B; EP1004650-B1; DE69824182-E.

16 Pérez Méndez M, Sanguino Otero J. Cholesteric Liquid-Crystal Copolyester, Poly[oxycarbonyl-1,4-phenylene- oxy - 1,4 terephthaloyl- oxy- 1,4-phenylenecarbonyloxy(1,2-dodecane)] [C34H36O8]n, Synthesized from Racemic Materials: Kinetics, Structure and Optical Characterization. International Journal of Engineering Research and Applications (IJERA) 2015; Vol. 5, Issue 7 (Part - 2): 48-62. ISSN: 2248-9622.

17 Perez-Mendez M, Fayos J, Blanch G P and Sánchez Cortés S. Biofunctionalization of Cholesteric Liquid-Crystal Helical Polymers. Nanocarriers. ENCYCLOPEDIA OF NANOSCIENCE.AND NANOTECHNOLOGY 2011, Volume 11: 547-580, Edited by H. S. Nalwa, ACS. American Scientific Publishers, ISBN:1-58883-160-4;

18 Pérez Méndez M, Hammouda B. SAXS and SANS investigation of synthetic cholesteric liquid-crystal polymers for biomedical applications. Journal of Materials Science and Engineering 2013; B 3 (2): 104-115.

19 Pérez Méndez M. Synthetic Cationic Cholesteric Liquid Crystal Polymers. In: Liquid Crystals - Recent Advancements in Fundamental and Device Technologies, Chapter 2. Dr. Pankaj Kumar Choudhury (Ed.): InTechOpen; 2018. DOI: 10.5772/intechopen.70995. https://www.intechopen.com/books/ liquid-crystals-recent-advancements-in- fundamental-and-device-technologie s/synthetic-cationic-cholesteric-liquid- crystal-polymers.

20 Freire F, Seco J M, Quiñoá E, Riguera R. The Prediction of the Absolute Stereochemistry of Primary and Secondary 1,2-Diols by 1H NMR Spectroscopy: Principles and Applications. Chem. Eur.J. 2005; Vol. 11; Issue 19: 5509-5522.

21 Arias S, Freire F, Quiñoá E, Riguera R.Nanospheres, Nanotubes, Toroids, and Gels with Controlled Macroscopic Chirality. Angew. Chem. Int. Edition 2014; 53, Issue 50: 13720– 13724.

22 Ute K, Hirose K, Kashimoto H, Hatada K, Vogl O. Haloaldehyde polymers. 51. Helix-sense reversal of isotactic chloral oligomers in solution. J. Am. Chem. Soc. 1991; 113: 6305- 6306, DOI: 10.1021/ja00016a076.

23 Ute K, Oka K, Okamoto Y, Hatada K, Xi F, Vogl O. Haloaldehyde Polymers LIII. Optical Resolution of Purely Isotactic Oligomers of Chloral: Optical Activity of the Chloral Oligomers Assuming One-Handed Helical Conformation in Solution. Polym. J. 1991; 23: 1419-1424.

24 Ute K, Hirose K, Kashimoto H, Nakayama H, Hatada K, O Vogl O. Helix-Inversion Equilibrium of Isotactic Chloral Oligomers in Solution. Polym. J. 1993; 25, N° 11: 1175- 1186.

25 Fayos J, Sánchez-Cortés S, Marco C, Pérez-Méndez M. Conformational analysis and molecular modeling of cholesteric liquid-crystal polyesters based on XRD, Raman and transition thermal analysis. J. Macromol.Sci.- Physics 2001, B40(3&4): 553-576. https://doi.org/10.1081/MB-100106177.

26 Perez-Mendez M, Marsal R, Garrido L, Martin-Pastor M. Self- Association and Stereoselectivity in a Chiral Liquid-Crystal Cholesteric Polymer Formed under Achiral Conditions. Macromolecules 2003; 36: 8049-8055.

27 Tian G, Lu Y, Novak B M. Helix- Sense Selective Polymerization of Carbodiimides: Building Permanently Optically Active Polymers from Achiral Monomers. J. Am. Chem. Soc. 2004; 126: 4082-4083.

28 Schlitzer D S, Novak B M. Trapped Kinetic States, Chiral Amplification and Molecular Chaperoning in Synthetic Polymers: Chiral Induction in Polyguanidines through Ion Pair Interactions. J. Am. Chem. Soc. 1998; 120, 9: 2196-2197.

29 Tang H Z, Lu Y, Tian G, Capracotta M D, Novak B M. Stable Helical Polyguanidines: Poly{N-(1-anthryl)-N'- [(R)-and/or (S)-3,7-dimethyloctyl] guanidines}. J. Am. Chem. Soc. 2004; 126: 3722-3723.

30 Materials Studio v.2021, Dassault Systemes BIOVIA, Cambridge, U.K.

31 Flory P J. Principles of Polymer Chemistry, Cornell University Press: Ithaca (1953): p. 400.

32 Boiko N, Shibaev V. Cholesteric polymer liquid crystals and their optical properties. International Journal of Polymeric Materials. 2000;45(3–4): 33 533-583

33 ATSAS 3.0: expanded functionality and new tools for small-angle scattering data analysis. Manalastas-Cantos K, Konarev P V, Hajizadeh N R, Kikhney A G, Petoukhov M V, Molodenskiy D S, Panjkovich A, Mertens H D T, Gruzinov A, Borges C, Jeffries C M, Svergun D I, Franke D. J. Appl. Cryst. 2021; 54: 343-355. DOI:10.1107/ S1600576720013412

34 Young R A, editor. The Rietveld method. IUCr Monograph on Crystallography; No. 5. Oxford: International Union of Crystallography/ Oxford University Press 1993. ISBN 36 0-19-855577-6.

35 McCusker L B, Von Dreele R B, Cox D E, Louër D, Scardie P. Rietveld refinement guidelines. J. Appl. Cryst. 1999; 32: 36-50.

5

High Precision Optical Wavefront Generation Using Liquid Crystal Spatial Light Modulator (LC-SLM)

1. Introduction

A spatial light modulator is a device that modulates the spatial distribution of light waves. Generally speaking, the spatial light modulator is composed of many independent units, which are arranged into one-dimensional or two-dimensional array structures in space. Each unit independently receives the control of optical signal or electrical signal, and changes the amplitude or intensity, phase, and polarization of light received in space. Because of the excellent properties of liquid crystal, liquid crystal spatial light modulator (LC-SLM) is widely used in adaptive optics [1], diffractive optical elements [2], optical testing [3], and so on [4].

According to the different addressing modes of the spatial light modulator, it can be divided into electrical addressing spatial light modulator (EA-SLM) and optical addressing spatial light modulator (OA-SLM). The electrically addressable spatial light modulator usually adds the signal to the corresponding unit through two groups of orthogonal grid electrodes on the SLM by means of progressive scanning. The input signal of the optically addressable spatial light modulator is an optical signal, which can convert the intensity distribution of writing light into charge distribution, refractive index distribution, and so on. In recent years, due to the rapid developments of liquid crystal display and VLSI technology and the abundance of liquid crystal materials, the application of electrically addressable liquid crystal spatial light modulators as wavefront correction devices in adaptive optics has attracted more and more attention. As a result, electrically addressable LC-SLM has great potential in realizing high-resolution wavefront control of optical systems.

The typical structure of a reflective LC-SLM is shown in Figure 1. It looks like a "sandwich" with three parts. The upper part is the covering glass with a transparent conductive film. The middle part is a liquid crystal layer containing thousands of liquid crystal molecules. And the bottom part is the silicon substrate containing discontinuous reflection pixels.

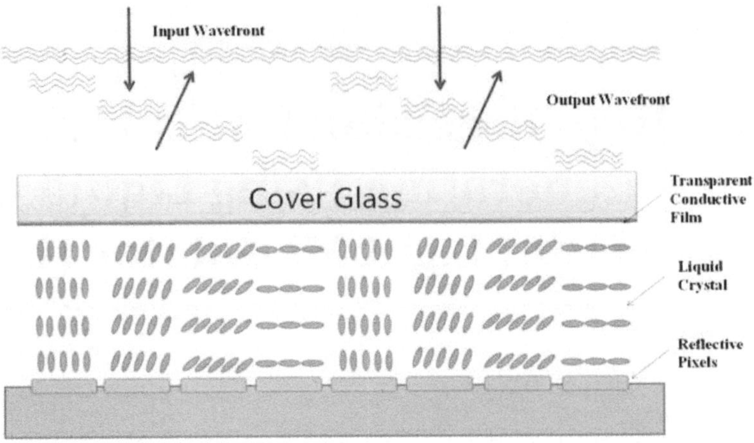

Figure 1. Typical structure of a reflective LC-SLM.

When the voltage is applied, the molecular structure of the liquid crystal will twist, resulting in the change of the birefringence coefficient of the liquid crystal. This electro-optic effect is called the electrically controlled birefringence effect. The electric field makes the liquid crystal molecules polarized and deflected, and changes the arrangement of liquid crystal molecules. With the increase of voltage, the liquid crystal molecules will break away from the intermolecular attraction and gradually incline along the electric field. When the threshold voltage is exceeded, except for the surface viscous force at the electrode substrate, other liquid crystal molecules will rearrange along the electric field direction. Different phase modulation can be generated by controlling the liquid crystal voltage, as shown in Figure 1. The phase delay between extraordinary light (e light) and ordinary light (o light) is shown in Eq. (1).

$$\delta = \frac{4\pi}{\lambda}\int_0^d (n_e(\theta) - n_o)dz \qquad (1)$$

where d is the thickness of the liquid crystal layer, and ne(θ) represents the refractive index of extraordinary light and can be expressed by Eq. (2).

$$n_e(\theta) = \frac{n_e \cdot n_o}{\sqrt{n_e \sin^2\theta + n_o \cos^2\theta}} \qquad (2)$$

where the deflection angle θ is related to the applied voltage and can be expressed by Eq. (3).

$$\theta = \begin{cases} 0 & V \leq V_C \\ \frac{\pi}{2} - 2\tan^{-1}\left\{\exp\left[-\left(\frac{V - V_C}{V_0}\right)\right]\right\} & V > V_C \end{cases} \quad (3)$$

where Vc is the initial voltage of the liquid crystal molecule when it starts to deflect, namely the threshold voltage. Therefore, the phase modulation produced by the electronic birefringence is related to the voltage applied at both ends of the liquid crystal layer.

It can be clearly seen from Eq. (3) that the relationship between the deflection angle and the applied voltage is nonlinear. As a result, the phase response of the LC-SLM is nonlinear, which needs to be calibrated accurately.

2. Nonlinear phase response calibration

2.1 Traditional methods to measure the phase modulation curve of LC-SLM

The phase response calibration is to measure the phase modulation curve with respect to the applied voltage (grayscale). In general, the measurement methods of phase modulation characteristics can be divided into two groups—the interference method and the diffraction method. Among them, interferometry mainly includes double-slit/hole interferometry [5, 6, 7], Twyman-Green interferometry [8, 9, 10, 11], Mach-Zehnder interferometry [12, 13, 14], and digital holographic interferometry [15, 16, 17]. The measurement of phase modulation characteristics of interferometry mainly depends on the displacement of fringe pattern, but the two beams have to travel a long path in the air before the interference, and the mechanical vibration, air turbulence, and other environmental factors will cause the change of their optical path difference, resulting in a large fluctuation in the acquisition of fringe pattern. Diffraction methods are mainly based on irradiance measurements of the diffraction pattern originated by phase holograms at their focal planes [18, 19, 20, 21]. Intensity transmission can well suppress the influence of environmental vibration and air turbulence, but from the perspective of phase extraction, the operation process of phase estimation is more complex. Here, we mainly introduce some commonly used calibration methods.

2.1.1 Diffraction-based methods

Figure 2 shows a typical configuration for phase modulation measurement based on the diffraction of the loaded phase hologram. After beam expansion and collimation, the polarized plane wave is divided into two beams by the beam splitter (BS). One beam of light is detected by the first photodiode (PD1), which is used for the correction of any power jitter. Another beam is reflected by LC-SLM where a binary grating is loaded. And the first diffraction order light of the grating is detected by the second photodiode (PD2). According to the Fourier optics theory [22], the diffraction efficiency of the binary grating is related to the phase difference of the two levels. As a result, by changing the phase difference of the loaded two-level grating, the phase modulation value can be calculated from Eq. (4) [19].

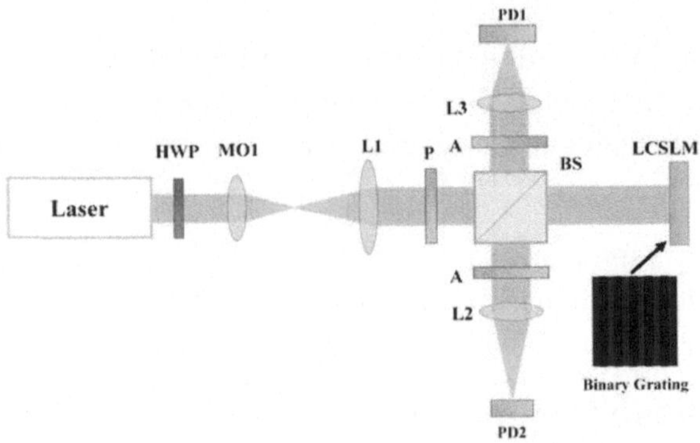

Figure 2. Schematic of the diffraction-based method by using a binary grating. HWP—half waveplate, MO—micro objective, L1, L2—lens, P—polarizer, A—analyzer, PD1, PD2—photodiode.

$$\delta = \sin^{-1}\frac{\pi}{2}\sqrt{\frac{P_2}{P_1}} \tag{4}$$

where P1 and P2 represent the intensity detected by the PD1 and PD2, respectively. Note that the optical system shown in Figure 2 is somewhat complex. To simplify the system, a circular grating (binary Fresnel phase lens) [20] was adopted instead of a linear grating. Due to the use of circular grating, the physical lens is no longer needed and the use of minimal optical elements allows a fast alignment of the experimental setup as shown in Figure 3. This compact configuration makes it suitable for in situ calibration for SLM.

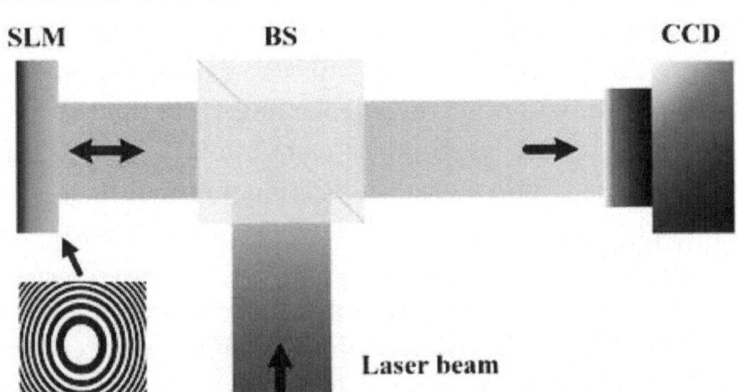

Figure 3. Simplified setup of the diffraction-based method by using a binary circular grating [20].

2.1.2 Interferometry-based methods

Different from the diffraction-based methods where the phase response is characterized by the first-order diffraction efficiency of the loaded binary grating, the interferometry-based methods utilized the movement of the interference fringe to calculate the phase shift value. The first commonly used interferometric method is the double-slit/hole interferometry whose optical setup is shown in Figure 4.

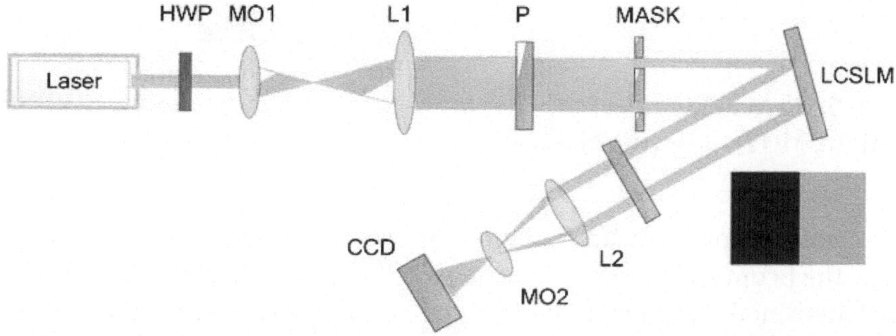

Figure 4. Optical setup of the double-slit interferometry. HWP—half wave plate, MO1, MO2—micro-objective, L1, L2—lens, P—polarizer.

After the laser passes through the half-wave plate (HWP), it forms a plane wave through the beam expansion collimation system composed of micro-objective lens (MO1) and lens (L1). Then, it needs to travel through the polarizer (P) and the mask with two holes (or slits) placed in front of the LC-SLM. At this time, the parallel light is divided into two beams by mask and incident on the target surface of LC-SLM, respectively. The gray image loaded into LC-SLM consists of two equal parts, one of which has a constant gray value of 0, and the other increases gradually from 0 to 255. Then, the two beams modulated by LC-SLM are focused by lens L2 and amplified by MO2. The interference fringes of the two beams are recorded by CCD. The double-slit/hole interferometry belongs to the common path interferometry, which is not easy to be interfered by the environmental turbulences, but the interference only occurs in the light transmission area of the mask, so the measured results can only reflect the modulation results of the local range of the target surface, and cannot accurately detect the phase modulation characteristics of the whole working surface. Recently, some researchers [23] used the SLM itself to generate the double holes so that the physical aperture with two holes is no longer needed. Since the calibration area is more easily adjusted, it can be used in different experimental conditions. As a matter of fact, if the double-slit/hole (mask) is replaced by a grating [24], the first-order diffracted beams can also interference after passing the lens. However, the zero-order light needs to be blocked out so as to get an interference fringe pattern with a good contrast.

Another commonly used method is the Twyman-Green interferometry, whose optical layout is shown in Figure 5.

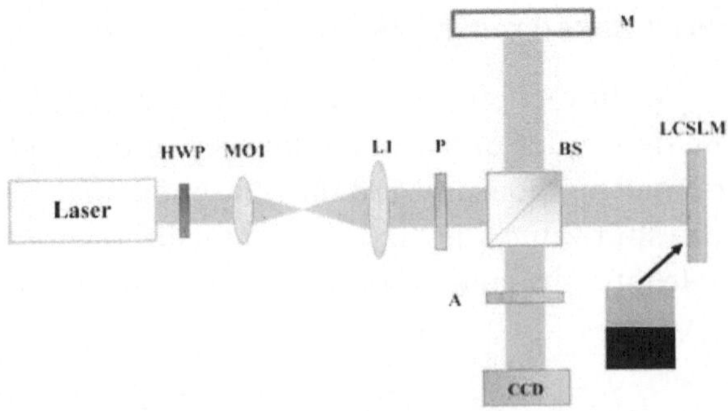

Figure 5. Optical layout of the Twyman-green interferometry. HWP—half waveplate, MO1—micro-objective, L1—lens, P—polarizer, A—analyzer, M—mirror.

After beam expansion and collimation, the plane wave is divided into two beams by the beam splitter (BS). One beam of light is perpendicular to LC-SLM and reflected after modulation by LC-SLM with loading grayscale image. The modulated light then interferes with the light reflected by the plane mirror (M). A CCD is used to record the interference fringes, and the phase modulation curve of LC-SLM is obtained by calculating the relative fringe movement over the fringe period. The results obtained by using the Twyman-Green interferometric method can detect the phase modulation characteristics of the whole working area. However, the two beams travel different paths before they can interfere. As a result, the method is greatly affected by the ambient vibration and air turbulence, which easily causes the fringe jitter and affects the measurement accuracy.

Recently, a radial shear interferometry was proposed by sending a Chinese high-order Taiji lens onto SLM [25]. The optical setup is shown in Figure 6. The method is realized by rotating multiple airy points, which are generated by the radial shear interference of high-order Chinese Taiji lens. As a result, the phase modulation value is related to the rotation angle of the captured image by CCD. However, the phase shift estimation is highly dependent on the accurate centroid location of the two Airy spots.

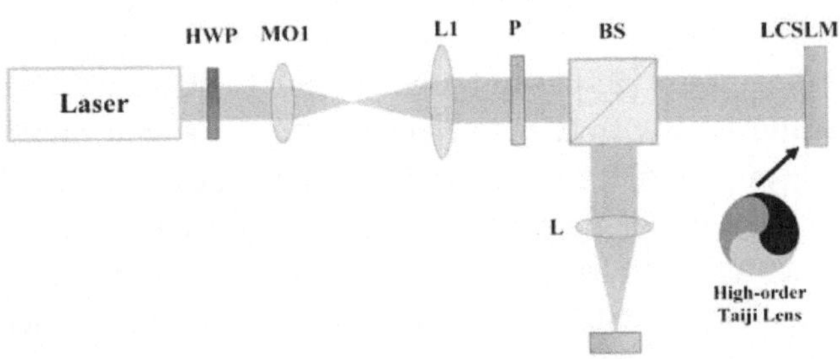

Figure 6. Optical setup of the radial shear interferometry [25]. HWP—half waveplate, MO1—micro-objective, L1, L2—lens, P—polarizer, A—analyzer.

2.2 Measuring the phase modulation curve of LC-SLM using the self-interference method

As mentioned above, the movement of the interference fringe is used to measure the phase modulation value in the interferometry-based methods. Different phase patterns loaded onto SLM will generate different kinds of interferograms. Inspired by the method in Ref. [26], we propose a self-interference method by using a diffraction grating [27]. The optical layout of the self-interference method is shown in Figure 7.

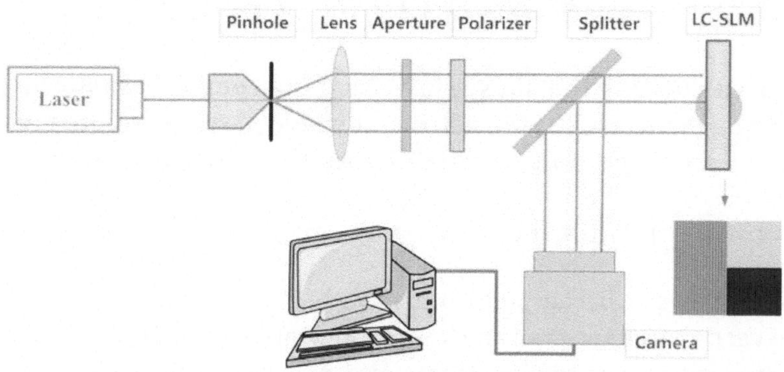

Figure 7. Optical layout of the self-interference method [27].

As it can be seen in Figure 7, the collimation beams perpendicularly strike the SLM and a beam splitter was used to deflect the reflected beams to the CCD plane. The combined gray pattern loaded on LC-SLM is divided into three parts. The left side is LC-SLM blazed grating with a period of 16 pixels, and the right side is divided into upper and lower parts. The lower part of the gray is zero and remains unchanged in the measurement process, which is called the reference part. In addition, the gray level of the upper part gradually increases from 0 to 255 in 8 steps, which is called the test part. After the reflection of LC-SLM, the first-order diffraction light and the left zero-order diffraction light of blazed grating interfere with LC-SLM at a certain distance, resulting in dislocation fringes. With the change of gray level in the right upper part, a series of fringe patterns with different shearing displacements could be obtained. The phase modulation value could then be calculated by using only one interferogram. As a result, the self-interference method can reduce the effect of environmental vibration or air turbulence and improve the measurement precision.

2.3 Phase calibration result

The captured fringe pattern is shown in Figure 8(a). Note that light is diffracted vertically from the sharp edge between the uniform grayscale zones, which causes unwanted effects in the fringe pattern. To reduce the diffraction effect, only a small part of the original fringe pattern (as shown in Figure 8(b)) was used to calculate the phase shift values. As shown in Figure 8(b), the fringes passing through the red scan line represented the measuring area, and the fringes passing through the blue scan line represented the reference area. The blue line reference area can be expressed as

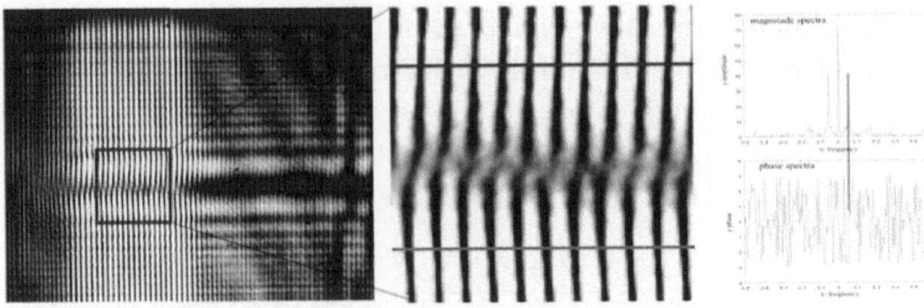

(a) (b) (c)

Figure 8. Calculating the phase shift using the Fourier transform: (a) the original fringe pattern, (b) the extracted fringe pattern denoted by reading rectangle, and (c) the FFT result of one row in (b).

$$i_1(x) = a(x) + b(x)\cos\left[2\pi f_x x + \varphi_0\right] \quad (5)$$

where a(x) is the background intensity, b(x) is the modulation depth, fx is the spatial carrier frequency of the x direction, φ0 is the initial phase of the blue reference area fringe. The red line measurement area can be expressed as

$$i_2(x) = a(x) + b(x)\cos\left[2\pi f_x x + \varphi_0 + \xi\right] \quad (6)$$

where ξ is the amount of phase shift between the measurement area and the reference area. The interference fringes given in Eqs. (5) and (6) are subjected to a Fourier transform:

$$F(f) = \int_{-\infty}^{\infty} i(x)\exp(-2\pi jfx)dx \quad (7)$$

where $j = \sqrt{-1}$. Then, we can extract the first-order spectrum and obtain its inverse Fourier transform:

$$i_{11}(x) = \int_{-\infty}^{\infty} F(f)\exp(2\pi jfx)df = c(x)\exp\left\{ j\left[2\pi f_x x + \varphi_0\right]\right\} \quad (8)$$

$$i_{21}(x) = \int_{-\infty}^{\infty} F(f)\exp(2\pi jfx)df = c(x)\exp\left\{ j\left[2\pi f_x x + \varphi_0 + \xi\right]\right\} \quad (9)$$

where $c(x) = \frac{1}{2}b(x)$

$$2\pi f_x x + \varphi_0 = \arctan \frac{\mathrm{Im}[i_{11}(x)]}{\mathrm{Re}\left[i_{11}(x)\right]} \tag{10}$$

$$2\pi f_x x + \varphi_0 + \xi = \arctan \frac{\mathrm{Im}[i_{21}(x)]}{\mathrm{Re}\left[i_{21}(x)\right]} \tag{11}$$

Finally, the phase shift ξ is calculated by subtracting Eq. (11) from (10):

$$\xi = \arctan \frac{\mathrm{Im}[i_{21}(x)]}{\mathrm{Re}\left[i_{21}(x)\right]} - \arctan \frac{\mathrm{Im}[i_{11}(x)]}{\mathrm{Re}\left[i_{11}(x)\right]} \tag{12}$$

The phase modulation was calculated by subtracting the phase of the two side lobes in the frequency domain shown in Figure 8(c). As a result, the relative phase shift of the upper and lower fringes can be obtained by Fourier transform phase analysis. In the actual calculation, 30 rows of data were used in one interferogram and 15 phase modulation values could be obtained. The final phase shift result is the average of these 15 values.

Figure 9 shows the phase modulation curves of a commercial SLM with different incident angles. It can be seen that the curves are almost coincident when the incident angle is less than 5 degrees, which indicates that the influence of the incident angle on the phase modulation is negligible when the angle is quite small. As the incident angle increases, the phase modulation curves become different. Particularly for the large gray level, their difference is very significant. As a result, the phase modulation depth decreases with the increase of the incident angle when it is larger than 10 degrees. To guarantee a good phase modulation capability, the incident angle is recommended to be less than 5 degrees in practical applications.

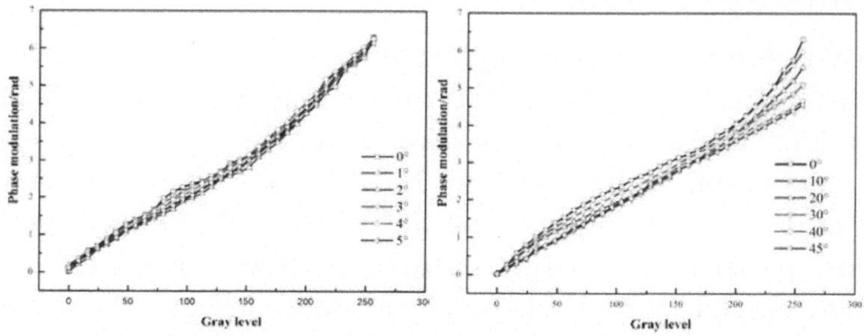

Figure 9. Phase modulation curves with different incident angles.

3. Static aberration measurement and compensation

Generally speaking, reflective LC-SLMs are more widely used in phase-only modulation, compared with transmissive LC-SLMs. The reason is that the reflective structure allows the incident light beam to travel the LC layer twice to obtain a double modulation depth. However, the static aberration of reflective silicon substrate or backplane, which is caused by the limitations in the polishing process at silicon foundries, leads to the uneven spatial response of SLM. To solve this problem and ensure the phase modulate precision of reflective LC-SLM, it is necessary to accurately measure and compensate the static aberration.

In recent years, several methods have been proposed to fulfill this task. These methods can also be divided into two categories—the diffraction-based methods and the interferometry-based methods. In the former category, the static aberration can be measured by applying a commercial wavefront sensor [28] or utilizing a phase retrieval technique [29, 30, 31]. The commercial wavefront sensor such as Shack-Hartman can only obtain a rough estimate of the static aberration. Compared with using a commercial wavefront sensor, the static aberration reconstructed by the phase retrieval technique is more accurate. However, the corresponding time consuming is higher due to the unavoidable iteration process, and the pixel cross talk effect impacts the accuracy of retrieval results as well. In practice, the methods of the latter category, interferometry-based methods, are more widely applied. In the latter category, typically a Michelson interferometer is used to capture the fringe pattern of the static aberration [32, 33, 34, 35, 36, 37, 38]. Furthermore, the fringe pattern can be demodulated to obtain the final true phase of aberration, by utilizing the phase-shift technology. Xun and Cohn [39] used the four-step method to demodulating the four interferograms with a phase step of $\pi/2\pi/2$. Later, Arias and Castaneda [38] measured the aberration of an LC-SLM by using Hariharan's five-step method. In their methods, the phase shift is introduced by a mechanical piezo-electric actuator, which is coupled to the reference mirror. Besides, Gongjian et al. [40] utilized the polarization phase-shifting technique to measure the static aberration of SLM. Here, we briefly introduce some recently reported methods.

3.1 Diffraction-based methods

A typically diffraction-based method is shown in Figure 10. By loading a random phase pattern onto SLM, the corresponding far-field diffraction pattern is captured by CCD. Then, an iterative phase retrieval technique is adopted to estimate the smooth aberration of the SLM [30]. Note that the pixel cross talk should be considered in the iteration process and this effect could be alleviated by using a random phase pattern with a larger feature size. Nevertheless, the measurement accuracy is still limited. Later, ptychography, as shown in Figure 11, was proposed to measure the static aberration by moving the SLM with a two-dimensional stage in a later direction [41]. Although the ptychography-based method can get a satisfactory result, the time efficiency is not very high. As a result, the interferometry-based methods are more widely used in the real application.

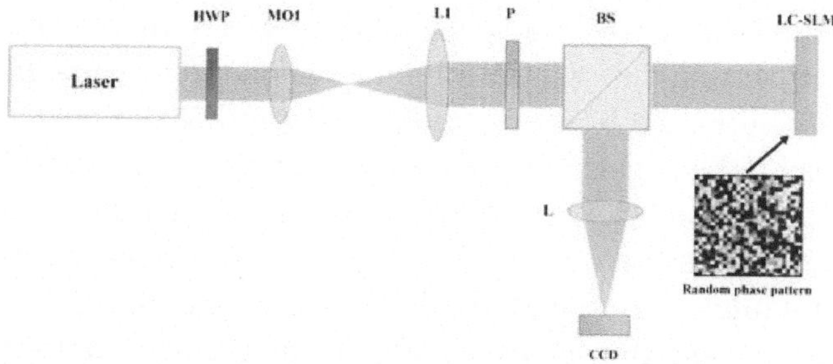

Figure 10. Optical setup of the static aberration measurement based on iterative phase retrieval [30]. HWP—half waveplate, MO1—micro-objective, L1, L2—lens, P—polarizer.

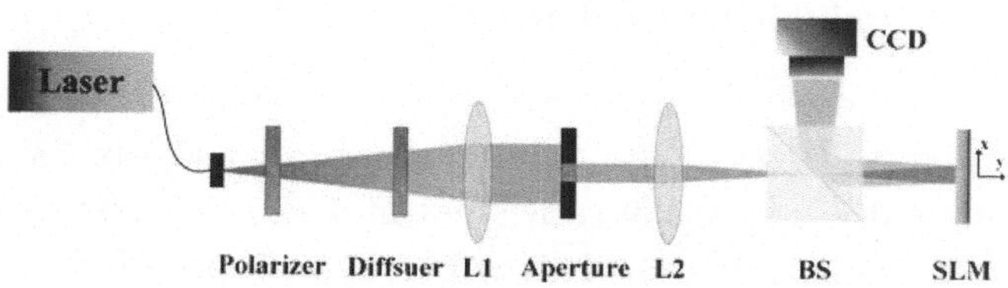

Figure 11. Optical layout of the static aberration measurement based on ptychography [41].

3.2 Interferometry-based method using random phase-shifting technique

Different from the traditional methods, here we introduce a novel interferometry-based method where the arbitrary phase shift is realized by the SLM itself [42]. And the phase is demodulated by a random phase-shifting technique. The configuration is quite simple and can be easily integrated into the optical system where SLM is used. The experimental schematic for this method is shown in Figure 12.

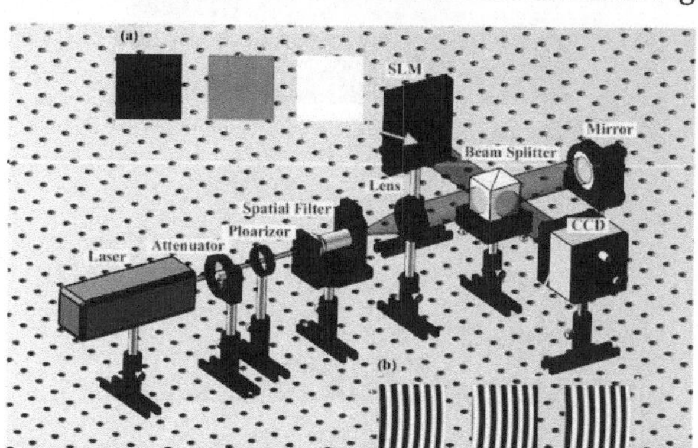

Figure 12. Experimental schematic for the static aberration measurement based on random phase-shifting interferometry [42]. (a) Constant grayscale patterns loaded on the SLM and (b) the corresponding fringe patterns captured by CCD.

A coherent light source with a wavelength of 632.8 nm is generated from the He-Ne laser. After passing through the attenuator, polarizer, spatial filter, and convex lens, a polarized collimated beam is obtained. Note that the polarization angle of the polarizer is set to be consistent with the modulation direction of SLM to ensure a pure phase modulation. The collimated beam is then divided into two parts by the beam splitter. One is the reference beam reflected by the mirror. The other is the test beam modulated by the SLM. These two beams interfere at the splitting surface of the splitter and the corresponding fringe pattern can be captured by the CCD. In addition, the phase shift is achieved by loading different images with the spatially consistent grayscale on the SLM. It should be noted that the mechanical phase shifter is no longer needed, which makes this configuration much simpler and more compact than traditionally used ones.

To show the validity of the introduced compensation method, the static aberration of a SLM (UPOLabs, HDSLM80R) is calibrated. Figure 13(a-d) show four images whose intensity is spatially consistent of 0, 63, 127, and 191, respectively. These images are loaded on the SLM to generate phase shifts, and then, CCD is able to capture four interference fringe patterns, as shown in Figure 13(e-h), respectively. These fringe patterns are demodulated by the VU factorization algorithm [43] and then unwrapped by the derivative Zernike polynomial fitting technique (DZPT) [44]. Figure 13(i) and (j) show the corresponding demodulated phase map and unwrapped phase map, respectively. Note that the unwrapped phase map is tilt-removed because the first three fitting coefficients of the Zernike polynomial are eliminated in the unwrapping process. Besides, Figure 13(k) shows the compensation image used to calibrate the static aberration. This compensation image is calculated by

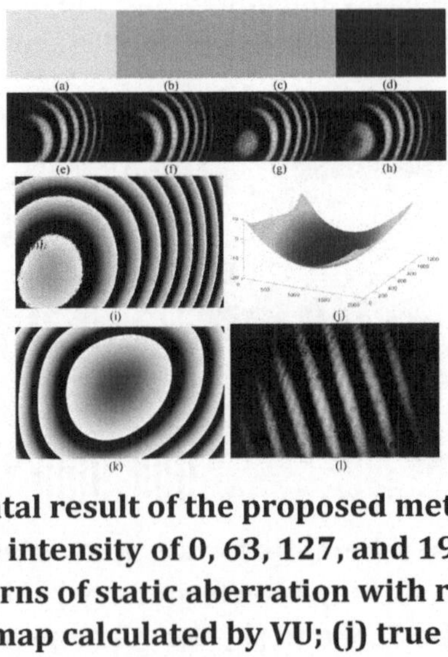

Figure 13. Experimental result of the proposed method; (a–d) four phase shift images with the intensity of 0, 63, 127, and 191, respectively; (e–h) four-step fringe patterns of static aberration with random phase shift; (i) demodulated phase map calculated by VU; (j) true phase map calculated by DZPT; (k) compensation phase map for SLM; (l) fringe pattern after compensation.

$$I(x,y) = \frac{255(\mathcal{W}(\psi(x,y)) + \pi)}{2\pi} \tag{13}$$

where ψ(x,y) represents the tilt removed true phase map, and I(x,y) is the compensation image needed to be loaded on SLM. After compensation, the interference image captured by CCD is shown in Figure 13(l). It can be seen from Figure 13(h) that the fringe pattern only contains some straight fringes corresponding to the tilt of the reference mirror or SLM, which verifies the effectiveness of the proposed method.

Furthermore, the validity of static aberration compensation results is verified by modulating a circular phase map. This phase map can be modeled as ϕcircular=10π(x2+y2) where x∈[−1,1] and y∈[−1200/1980,1200/1980], and its wrapped phase map is shown in Figure 14(a). Without static aberration compensation, the circular phase map is firstly modulated by loading its wrapped phase maps on SLM directly. It should be mentioned that the phase shift is realized by wrapping ϕcircular +0.5π, ϕcircular+π, and ϕcircular+1.5π, respectively. Figure 14(b) shows one of the fringe patterns captured by CCD, and the corresponding demodulated phase map is shown in Figure 14(c). It can be seen from Figure 14(b) and (c) that the phase modulated by SLM has large distortion since the demodulated result deviates from a spherical phase profile. For comparison, the circular phase map is modulated after static aberration compensation as well. To do this, the wrapped phase maps W(ψ+ϕcircular), W(ψ+ϕcircular+0.5π), W(ψ+ϕcircular+π), and W(ψ+ϕcircular+1.5π) are generated and loaded on SLM. Figure 14(d) and (e) show one of the wrapped phase maps and its corresponding fringe pattern captured by CCD, respectively, and the demodulated phase map can be seen in Figure 14(f). As it can be seen from Figure 14(e) and (f), a pretty good phase map that is very close to the spherical phase profile is obtained. Comparing the results before and after static aberration compensation, it is evident that the static aberration compensation significantly improves the quality of the reconstructed spherical wavefront, which further demonstrates that the back panel curvature (static aberration) has been compensated effectively.

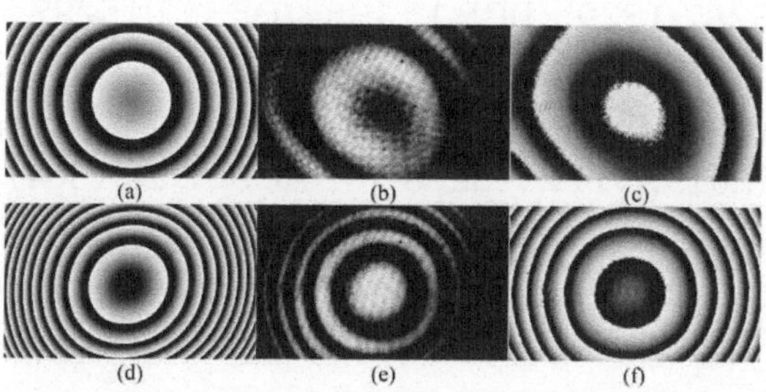

Figure 14. Spherical wavefront reconstruction results (a–c) without and (d–f) with static aberration compensation; (a) and (d) are grayscale images to be loaded on SLM; (b) and (e) are fringe patterns captured by CCD; (c) and (f) are demodulated phase maps by VU method [43].

4. Conclusion and discussion

To improve the phase control accuracy of LC-SLM, two factors, nonlinear response, and static aberration are comprehensively studied. A phase calibration method based on the self-generated grating by LC-SLM is introduced. Because of the common path configuration, the self-interference method can accurately obtain a phase modulation curve. Besides, a random phase-shifting interferometry is introduced to measure the static aberration of a reflective SLM. With the help of phase calibration and static aberration compensation, the quality of the reconstructed wavefront by LC-SLM is greatly improved. However, other factors (such as pixel cross talk, internal Fabry–Perot cavity, fill factor, bit depth, phase flicker) also affect the phase control accuracy. A plenty of researchers have proposed some method to compensate the effect of the pixel cross talk [45, 46, 47, 48], the phase flicker [49, 50, 51, 52, 53]. Nevertheless, most previous works mainly focused on the compensation of one factor. Recently, Pushkina [54] established a comprehensive model to compensate the effect of pixel cross talk, the back panel curvature (static aberration), and the internal Fabry-Perot cavity simultaneously. As a result, the performance of LC-SLM has been substantially improved. In general, different types of SLMs may have different optimized models. How to establish the optimal model for a specific LC-SLM by considering all the factors; and what is the best way to calibrate it? These are very interesting topics that need further research.

References

1. Chao L, Mingliang X, Quanqun M, et al. High-precision open-loop adaptive optics system based on LC-SLM. Optics Express. 2009;17(13):10774-10781. DOI: 10.1364/OE.17.010774
2. Kuang Z, Perrie W, Leach J, Sharp M, Edwardson SP, Padgett M, et al. High throughput diffractive multi-beam femtosecond laser processing using a spatial light modulator. Applied Surface Science. 2008;255(5p1):2284-2289. DOI: 10.1016/j.apsusc.2008.07.091
3. Justo A, Vicente D, Zbigniew J, et al. Measurement and compensation of optical aberrations using a single spatial light modulator. Optics Express. 2007;15(13):15287-15292. DOI: 10.1364/OE.15.015287
4. Lazarev G, Chen P-J, Strauss J, Fontaine N, Forbes A. Beyond the display: Phase-only liquid crystal on Silicon devices and their applications in photonics [Invited]. Optics Express. 2019;27(11):16206-16249. DOI: 10.1364/oe.27.016206
5. Soutar C, Monroe SE, Knopp J. Complex characterization of the Epson liquid crystal television. Proceedings of SPIE—The International Society for Optical Engineering. 1959;1993:269-277. DOI: 10.1117/12.160293
6. Duffey JN, Jones BK, Loudin JA, Booth JJ. Optical characterization of the InFocus TVT-6000 liquid crystal television (LCTV) using custom drive electronics. Proceedings of SPIE—The International Society for Optical Engineering. 1995;275(4):2349-2358. DOI: 10.1117/12.205778

7. Bergeron A, Gauvin J, Gagnon F, Gingras D, Doucet M. Phase calibration and applications of a liquid-crystal spatial light modulator. Applied Optics. 1995;34(23):5133-5139. DOI: 10.1364/AO.34.005133

8. Zhang H, Zhang J, Wu L. Evaluation of phase-only liquid crystal spatial light modulator for phase modulation performance using a Twyman-Green interferometer. Measurement Science & Technology. 2007;18(6):1724-1728. DOI: 10.1088/0957-0233/18/6/S09

9. Wu Y, Nie J, Shao L. Method to measure the phase modulation characteristics of a liquid crystal spatial light modulator. Applied Optics. 2016;55(31):8676-8682. DOI: 10.1364/AO.55.008676

10. Dai Y, Antonello J, Booth MJ. Calibration of a phase-only spatial light modulator for both phase and retardance modulation. Optics Express. 2019;27(13):17912-17926. DOI: 10.1364/OE.27.017912

11. Mukhopadhyay S, Sarkar S, Bhattacharya K, Hazra L. Polarization phase shifting interferometric technique for phase calibration of a reflective phase spatial light modulator. Optical Engineering. 2013;52(3):5602. DOI: 10.1117/1.OE.52.3.035602

12. Jeffrey A. Davis. Transmission and phase measurement for polarization eigenvectors in twisted-nematic liquid crystal spatial light modulators. Optical Engineering. 1998;37(11):3048-3052. DOI: 10.1117/1.601976

13. Reichelt S. Spatially resolved phase-response calibration of liquid-crystal-based spatial light modulators. Applied Optics. 2013;52(12):2610-2618. DOI: 10.1364/ao.52.002610

14. Wang H, Dong Z, Fan F, Feng Y, Lou Y, Jiang X. Characterization of spatial light modulator based on the phase in Fourier domain of the hologram and its applications in coherent imaging. Applied Sciences. 2018;8(7):1146. DOI: 10.3390/app8071146

15. Yang L, Xia J, Chang C, Zhang X, Yang Z, Chen J. Nonlinear dynamic phase response calibration by digital holographic microscopy. Applied Optics. 2015;54(25):7799-7806. DOI: 10.1364/AO.54.007799

16. Li R, Gao Y, Cao L. In situ calibration for a phase-only spatial light modulator based on digital holography. Optical Engineering. 2020;59(5):053101. DOI: 10.1117/1.OE.59.5.053101

17. Tiwari V, Gautam SK, Naik DN, Singh RK, Bisht NS. Characterization of a spatial light modulator using polarization-sensitive digital holography. Applied Optics. 2020;59(7):2024-2030. DOI: 10.1364/AO.380572

18. Engstrom D, Milewski G, Bengtsson J, Galt S. Diffraction-based determination of the phase modulation for general spatial light modulators. Applied Optics. 2006;45(28):7195-7204. DOI: 10.1364/AO.45.007195

19. Zhang Z, Yang H, Robertson B, Redmond M, Pivnenko M, Collings N, et al. Diffraction based phase compensation method for phase-only liquid crystal

on silicon devices in operation. Applied Optics. 2012;51(17):3837-3846. DOI: 10.1364/AO.51.003837

20. Mendoza-Yero O, Mínguez-Vega G, Martínez-León L, Carbonell-Leal M, Fernández-Alonso M, Doate-Buendía C, et al. Diffraction-based phase calibration of spatial light modulators with binary phase fresnel lenses. Journal of Display Technology. 2016;12(10):1027-1032. DOI: 10.1109/JDT.2016.2580902
21. Ronzitti E, Guillon M, de Sars V, Emiliani V. LCoS nematic SLM characterization and modeling for diffraction efficiency optimization, zero and ghost orders suppression. Optics Express. 2012;20(16):17843-17855. DOI: 10.1364/OE.20.017843
22. Goodman JW. Introduction to Fourier Optics. 3rd ed. Greenwoood Village: Roberts & Company Publishers; 2005
23. Li Y, Li Y, Zhang J, Li Y, Wang Z, Yin C, et al. Generalized phase calibration method of liquid crystal spatial light modulator with absolute reference system of obnoxious background light. Optics and Lasers in Engineering. 2020;132:106132. DOI: 10.1016/j.optlaseng.2020.106132
24. Bondareva AP, Cheremkhin PA, Evtikhiev NN, Krasnov VV, Starikov RS, Starikov SN. Measurement of characteristics and phase modulation accuracy increase of LC SLM \"HoloEye PLUTO VIS\". Journal of Physics Conference. 2014;536:012011. DOI: 10.1088/1742-6596/536/1/012011
25. Yang Y, Li Y, Li Y, Zhang J, Zhang Y, Liu D, et al. Calibration of phase liquid crystal spatial light modulators using high-order Taiji radial-shearing interferometry. Laser Physics. 2020;30(2):025003. DOI: 10.1088/1555-6611/ab5e22
26. Fuentes JLM, Fernández EJ, Prieto PM, Artal P. Interferometric method for phase calibration in liquid crystal spatial light modulators using a self-generated diffraction-grating. Optics Express. 2016;24(13):14159-14171. DOI: 10.1364/OE.24.014159
27. Zhao Z, Xiao Z, Zhuang Y, Zhang H, Zhao H. An interferometric method for local phase modulation calibration of LC-SLM using self-generated phase grating. Review of Scientific Instruments. 2018;89(8):083116. DOI: 10.1063/1.5031938
28. López-Quesada C, Andilla J, Martín-Badosa E. Correction of aberration in holographic optical tweezers using a Shack-Hartmann sensor. Applied Optics. 2009;48(6):1084-1090. DOI: 10.1364/AO.48.001084
29. Hart NW. Characterizing static aberrations in liquid crystal spatial light modulators using phase retrieval. Optical Engineering. 2007;46(8):086601. DOI: 10.1117/1.2767258
30. Xiao C, Zhiguang S, Xiaotian C, et al. Backplane aberration calibration of spatial light modulators using a phase-retrieval algorithm. Applied Optics 2016;55(31):8916-8924. DOI: 10.1364/ao.55.008916
31. Jesacher A, Schwaighofer A, Fürhapter S, Maurer C, Bernet S, Ritsch-Marte M Wavefront correction of spatial light modulators using an optical vortex image Optics Express. 2007;15(9):5801-5808. DOI: 10.1364/OE.15.005801

32. Harriman JL, Linnenberger A, Serati SA, Gonglewski JD, Gruneisen MT, Giles MK. Improving spatial light modulator performance through phase compensation. Proceedings of SPIE—The International Society for Optical Engineering. 2004;5553:58-67. DOI: 10.1117/12.563394

33. Zeng Z, Li Z, Fang F, Zhang X. Phase compensation of the non-uniformity of the liquid crystal on silicon spatial light modulator at pixel level. Sensors. 2021;21(3):967. DOI: 10.3390/s21030967

34. He A, Quan C. Wavefront correction for spatial nonuniformity of the liquid crystal on silicon based spatial light modulator. Optics and Lasers in Engineering. 2019;121:377-388. DOI: 10.1016/j.optlaseng.2019.05.010

35. Xu J, Qin S, Liu C, Fu S, Liu D. Precise calibration of spatial phase response nonuniformity arising in liquid crystal on silicon. Optics Letters. 2018;43(12):2993-2996. DOI: 10.1364/OL.43.002993

36. Szatkowski M, Popiołek Masajada A, Masajada J. Optical vortex trajectory as a merit function for spatial light modulator correction. Optics and Lasers in Engineering. 2019;118:1-6. DOI: 10.1016/j.optlaseng.2019.01.014

37. Sheng Y, Zhou C, Yu C, Lai X, Wei K, Fan S, et al. Correction for the inherent aberration of liquid crystal spatial light modulator. Holography, Diffractive Optics, and Applications. 2018;10818:1081827. DOI: 10.1117/12.2502217

38. Arias A, Castaneda R. Estimation and Compensation of aberrations in Spatial Light Modulators. Journal of Physics: Conference Series. 2011;274(1):012060. DOI: 10.1088/1742-6596/274/1/012060

39. Xun X, Cohn RW. Phase calibration of spatially nonuniform spatial light modulators. Applied Optics. 2004;43(35):6400. DOI: 10.1364/AO.43.006400

40. Gongjian Z, Man Z, Yang Z. Phase modulation characteristics of spatial light modulator and the system for its calibration. Journal of Electrical Engineering. 2018;6(4):193-205. DOI: 10.17265/2328-2223/2018.04.001

41. McDermott S, Li P, Williams G, Maiden A. Characterizing a spatial light modulator using ptychography. Optics Letters. 2017;42(3):371-374. DOI: 10.1364/OL.42.000371

42. Li J, Zhao Z, Fan C, Du Y, Zhou M, Zhang X, et al., editors. Characterizing static aberration in reflective liquid crystal spatial light modulators (LC-SLM) using random phase shifting interferometry. In: 2021 International Conference of Optical Imaging and Measurement (ICOIM); 27-29 Aug Xi'an. New York: IEEE; 2021. 2021. DOI: 10.1109/ICOIM52180.2021

43. Escobar MA, Estrada JC, Vargas J. Phase-shifting VU factorization for interferometry. Optics and Lasers in Engineering. 2020;124:105797. DOI: 10.1016/j.optlaseng.2019.105797

44. Zhao Z, Zhao H, Zhang L, Gao F, Qin Y, Du H. 2D phase unwrapping algorithm for interferometric applications based on derivative Zernike polynomial fitting technique. Measurement Science and Technology. 2015;26(1):017001. DOI: 10.1088/0957-0233/26/1/017001

45. Jz A, Hla B, Xu LA. Reducing the crosstalk effect in phase-only spatial light modulators based on double-phase method. Optics Communications. 2020;465:125595. DOI: 10.1016/j.optcom.2020.125595

46. Simon M, Monika R-M, Gregor T. Model-based compensation of pixel crosstalk in liquid crystal spatial light modulators. Optics Express. 2019;27(18):25046-25063. DOI: 10.1364/OE.27.025046

47. Guesmi M, Žídek K. Calibration of the pixel crosstalk in spatial light modulators for 4f pulse shaping. Applied Optics. 2021;60(25):7648-7652. DOI: 10.1364/AO.434309

48. Chen Q, Shen X, Cheng Y, Liu J, Cai J, Liu Y. A crosstalk-reduced method of complex fields encoding using a single phase-only spatial light modulator. Optik. 2021;228:166190. DOI: 10.1016/j.ijleo.2020.166190

49. Márquez A, Martínez-Guardiola FJ, Francés J, Gallego S, Pascual I, Beléndez A. Combining average molecular tilt and flicker for management of depolarized light in parallel-aligned liquid crystal devices for broadband and wide-angle illumination. Optics Express. 2019;27(4):5238-5252. DOI: 10.1364/oe.27.005238

50. Moreno, Lizana, Márquez, Iemmi, Fernández. Time fluctuations of the phase modulation in a liquid crystal on silicon display: Characterization and effects in diffractive optics. Optics Express. 2008;16(21):16711-16722. DOI: 10.1364/OE.16.016711

51. Nobukawa T, Katano Y, Muroi T, Kinoshita N, Ishii N. Reduction of spatio-temporal phase fluctuation in a spatial light modulator using linear phase superimposition. OSA Continuum. 2021;4(6):1846-1858. DOI: 10.1364/OSAC.426628

52. Yang Z, Wu S, Nie J, Yang H. Uncertainty in the phase flicker measurement for the liquid crystal on silicon devices. Photonics. 2021;8(8):307. DOI: 10.3390/photonics8080307

53. Zheng M, Chen S, Liu B, Weng Z, Li Z. Fast measurement of the phase flicker of a digitally addressable LCoS-SLM. Optik. 2021;242:167270. DOI: 10.1016/j.ijleo.2021.167270

54. Pushkina A, Filho J, Maltese G, Lvovsky A. Comprehensive model and performance optimization of phase-only spatial light modulators. Measurement Science and Technology. 2020;31:125202. DOI: 10.1088/1361-6501/aba56b

6

The LCD Interfacing and Programming

1. Introduction

LCD stands for Liquid-Crystal Display. From the name, we know that this equipment is used as a display. Many types of electronic equipment (Laptop, Digital Voltmeter, Mobile Phone, ATM-Automatic Teller Machine, PC-Personal Computer, Compact Disk Player, Digital Thermometer, Clothe Washer, Flat TV-Television, etc.) has used this equipment to display images, Alpha Numeric information, and video. LCD is electronic equipment (display) made from liquid-crystal layers which operate if it gets varied voltage/currents which change the optical properties (light modulating properties) of this display; which causes this tool to continue or block the light through it. LCD uses its own light source known as Back Plane Light (BPL). Compared to other types of display such as CRT (Cathode Ray Tube), LED (Light Emitting Diode), plasma gas, LCD consumes smaller power, this is because it does not produce light, but this tool displays images with the principle of inhibiting or forwarding light; the changing of the optical properties of the LCD when it gets a varying voltage/current on the LCD layer and the existence of a light source (BPL) causes this equipment to display the object/character that can be seen by our eyes. At present, there are two types of LCDs known in general, namely,

- ❖ Passive matrix
- ❖ Active matrix

Matrix active is often referred to as AMLCD (Active Matrix LCD). Some people call AMLCD with TFT (thin-film transistor) this is due to a transistor that is useful as a large controller of the current through the intersection of pixels. With the use of transistors as controlling currents, controlling the current through intersection becomes faster which ultimately increases the refresh time of this tool which then produces subtler images compared to the LCD-passive matrix [1, 2].

The manufacture or the fabrication of AMLCD is far more complicated than the fabrication of passive matrix LCD which causes its price to be more expensive than the passive matrix LCD [2]. The Microcontroller System 51-MCS51 (Microcontroller 8051 family) can also be connected with the LCD, where the LCD is used as a display. On this occasion, we will discuss LCDs commonly used in various microcontroller applications namely LCD 2 × 16; 2 rows with 16 characters/row. The LCD module basically consists of two units namely the LCD itself and the controller. The LCD module is mounted and pinned to the controller and then the microcontroller is connected to the LCD controller.

1.1 2 × 16 LCD

There are various types of LCDs commonly used in various microcontroller applications, namely 1 × 16, 2 × 16, 4 × 20 [3]. The LCD type is known by the number of lines and the number of characters per line it has. LCD with type 1 × 16 means, this LCD can only display 1 line of characters, and can only display 16 alphanumeric characters. Likewise, LCD with types 2 × 16 and 4 × 20 each has 2 and 4 consecutive lines, and the first type can only display 16 alphanumeric characters/line, while the last type is capable of displaying 20 alphanumeric characters/line. On this occasion we will only discuss the connection/the usage of the second type of LCD, type 2 × 16 (2 lines × 16 characters/line); by understanding the 2 × 16 type LCD, then you can easily read and learn the manual of other types to connect them to MCS51 [4].

Figure 1 below shows the physical appearance of this display,

Figure 1. 2 × 16 LCD physical picture [5].

From Figure 1 it can be seen that the number of lines of the LCD is 2. The first line contains the words "this is a 2 × 16" while the second line contains the words " Line LCD Display" [5].

1.2 LCD controller

As previously mentioned above to build the LCD module, the LCD (M1632) is mounted on and pinned to the HD44780 controller. To use the LCD module in MCS-51 the controller is connected to the MCS-51 (Figure 2) [3, 6].

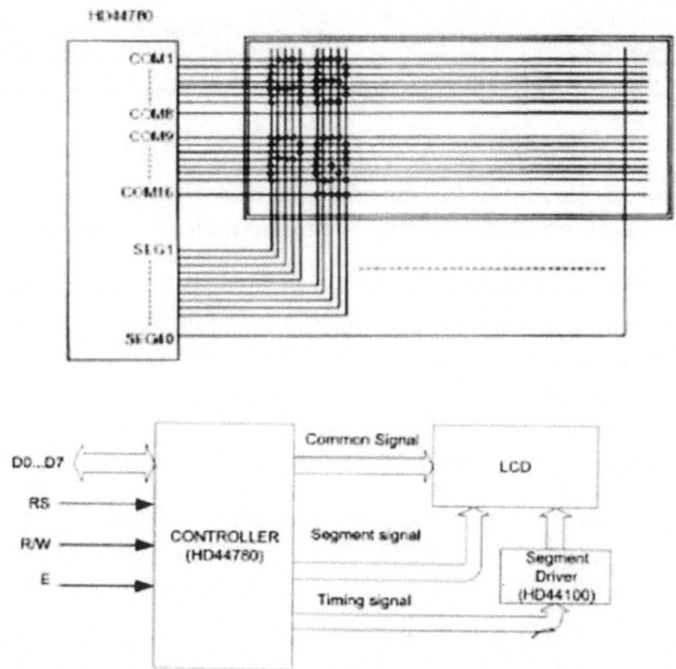

Figure 2. Block diagram of 2 × 16 LCD module [3, 6].

The LCD controller can be used for various types of LCDs with various numbers of characters per line; from 8 to 80 characters per line and the number of lines between 1 and 4 lines [4]. The controller has DISPLAY DATA RAM (DDRAM) in it with a memory map as shown in Figure 3 below [7].

00	01	02	03	04	05	06	07	08	09	0A	0B	0C	0D	0E	0F	10	11	12	13	14	15	16	17	18	19	1A	1B	1C	1D	1E	1F	20	21	22	23	24	25	26	27
40	41	42	43	44	45	46	47	48	49	4A	4B	4C	4D	4E	4F	50	51	52	53	54	55	56	57	58	59	5A	5B	5C	5D	5E	5F	60	61	62	63	64	65	66	67

Figure 3. 2 × 16 LCD DDRAM- memory map [6, 7].

From Figure 3 above we can see that the DDRAM controller should have a memory map of 104 bytes (00H to 67H = 68H = 104D), but the location that can be used is only 80 bytes 00H-27H (40 bytes) and 40H-67H (40 bytes); Note that the 28H-3FH address (24 location) is not visible.

Each place on the LCD has an address that corresponds to the address on the DDRAM on the LCD controller) see Figures 3 and 4 for more details. The DDRAM memory map on the 2 × 16 bit LCD controller that can be used for LCD with type M1632 is 00H-0FH for the first line (16 addresses) and 40H-4FH (16 addresses) for the second line (so the total address location 32 address locations), although as shown in Figure 3 DDRAM memory maps of the LCD controller basically amounted to 80 locations.

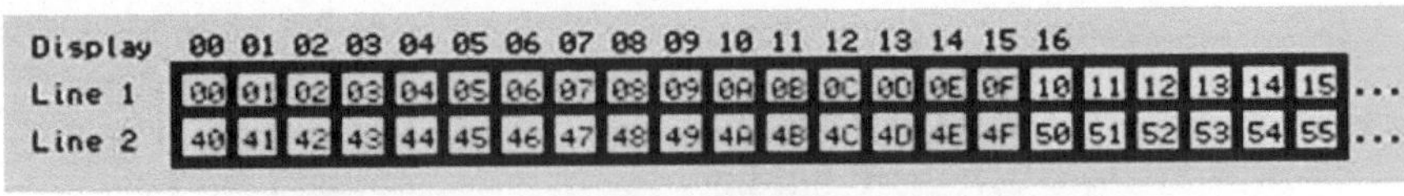

Figure 4. 2 × 16 LCD DDRAM address [8].

However, the display looks just having 00H-0FH addresses for the first line and 40H-4FH addresses for the second line as shown in Figure 4 above [8].

LCD connected to this controller will adjust itself to the memory map of this DDRAM controller; each location on the LCD will take 1 DDRAM address on the controller. Because we use 2 × 16 type LCD, the first line of the LCD will take the location of the 00H-0FH addresses and the second line will take the 40H-4FH addresses of the controller DDRAM; so neither the addresses of the 10H-27H on the first line or the addresses of the 50H-67H on the second line on DDRAM is used.

To be able to display a character on the first line of the LCD, we must provide written instructions (80h + DDRAM address where our character is to be displayed on the first line) in the Instruction Register-IR and then followed by writing the ASCII code of the character or address of the character stored on the CGROM or CGRAM on the LCD controller data register, as well as to display characters in the second row we must provide written instructions (C0H + DDRAM address where our character to be displayed on the second line) in the Instructions Register-IR and then followed by writing the ASCII code or address of the character on CGROM or CGRAM on the LCD controller data register.

The characters to be displayed can be in the form of American Standard Code For Information Interchange-ASCII code, characters originating from CGROM (various characters are prepared by the controller manufacturer or producer and are permanently stored in it), and from the character generator RAM-CGRAM (special place on CGROM with the addresses 00000000B-00000111B (consecutive eight locations) prepared for storing characters specifically to be displayed by the user/custom use); the total address of the CGROM with the CGRAM address in it and allocated specially for custom use is 256D (00000000B-111111111B) locations see Figure 5 below; please do not confuse CGROM/CGRAM address with DDRAM address they are two completely different and independent memory locations. The character stored on the CGRAM address is temporary (volatile) and must be stored in it before retrieve and displayed it on the LCD; the data stored in CGRAM will disappear if the source of electrical voltage is cut off; in contrast, the characters stored in the CGROM is permanently reserved in it.

Figure 5. CGRAM [9].

As mentioned above, to display a character (ASCII) you want to show on the LCD, you need to send the ASCII code to the LCD controller data register-DR. For characters from CGROM and CGRAM we only need to send the address of the character where the character is stored; unlike the character of the ASCII code, we must write the ASCII code of the character we want to display on the LCD controller data register to display it. For special characters stored on CGRAM, one must first save the special character at the CGRAM address (prepared 64 addresses, namely addresses 0–63); A special character with a size of 5 × 8 (5 columns × 8 lines) requires eight consecutive addresses to store it, so the total special characters that can be saved or stored on the CGRAM addresses are only eight (8) characters. To be able to save a special character at the first CGRAM address we must send or write 40H instruction to the Instruction Register-IR followed by writing eight consecutive bytes of the data in the Data Register-DR to save the pattern/image of a special character that you want to display on the LCD [9, 10].

We can easily connect this LCD module (LCD + controller) with MCS51, and we do not need any additional electronic equipment as the interface between MCS51 and it; This is because this LCD works with the TTL logic level voltage—Transistor-Transistor Logic.

Based on 2 × 16 LCD data sheets of the Hitachi 44780 trademark we can connect it with MCS-51 as shown in Figure 6 [11].

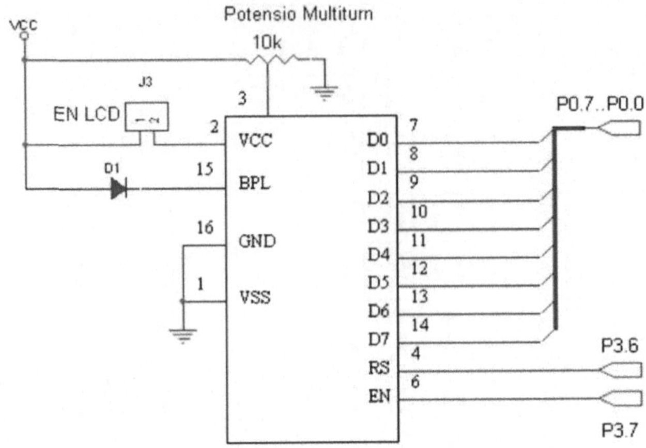

Figure 6. Connection of 2 × 16 LCD module (8 bit mode) with MCS51 [11].

From Figure 6 above it can be seen clearly that this display has 16 Pins that can be grouped into 4 as shown in Table 1.

Table 1. Pin function of 2 × 16 LCD.

Pin	Name Pin	Function
1, 2, dan 16	VSS, VCC, GND	Power supply
7–14	D0-D7	Data bus
4, 5, and 6	RS, R/$\overline{W}$, and $\overline{EN}$	Control
3 and 15	VEE and BPL	BPL

The voltage source of this display is +5 V connected to Pin 2 (VCC) and GND power supply connected to Pin 1 (VSS) and Pin 16 (GND); Pin 1 (VSS) and Pin 16 (GND) are combined together and connected to the GND of the power supply.

Pins 7–14 (8 Pins) of the display function as a channel to transmit either data or instruction with a channel width of 1 byte (D0-D7) between the display and MCS51. In Figure 6, it can be seen that each Pin connected to the data bus (D0-D7) of MCS51 in this case P0 (80h); P0.0-P0.7 MCS-51 connected to D0-D7 of the LCD.

Pins 4–6 are used to control the performance of the display. Pin 4 (Register Select-RS) is in charge of selecting one of the 2 display registers. If RS is given logic

0 then the selected register is the Instruction Register-IR, otherwise, if RS is given logic 1 then the selected register is the Data Register-DR. The implication of this selection is the meaning of the signal sent down through the data bus (D0-D7), if RS = 0, then the signal sent from the MCS-51 to the LCD is an instruction; usually used to configure the LCD, otherwise if RS = 1 then the data sent from the MCS-51 to the LCD (D0-D7) is the data (object or character) you want to display on the LCD. From Figure 6 Pin 4 (RS) is connected to Pin 16 (P3.6/W¯) of MCS-51 with the address (B6H).

Pin 5 (R/W¯)) of the LCD does not appear in Figure 6 is used for read/write operations. If Pin 5 is given logic 1, the operation is a read operation; reading the data from the LCD. Data will be copied from the LCD data register to MCS-51 via the data bus (D0-D7), namely Pins 7–14 of the LCD. Conversely, if Pin 5 is given a voltage with logical 0 then the operation is a write operation; the signal will be sent from the MCS51 to LCD through the LCD Pins (Pins 7–14); The signal sent can be in the form of data or instructions depending on the logic level input to the Register Select-RS Pin, as described above before if RS = 0 then the signal sent is an instruction, vice versa if the RS = 1 then the signal sent/written is the data you want to display. Usually, Pin 5 of the LCD is connected with the power supply GND, because we will never read data from the LCD data register, but only send instructions for the LCD work configuration or the data you want to display on the LCD.

Pin 6 of the LCD (EN ¯) is a Pin used to enable the LCD. The LCD will be enabled with the entry of changes in the signal level from high (1) to low (0) on Pin 6. If Pin 6 gets the voltage of logic level either 1 or 0 then the LCD will be disabled; it will only be enabled when there is a change of the voltage level in Pin 6 from high logic level to low logic level for more than 1000 microseconds (1 millisecond), and we can send either instruction or data to processed during that enable time of Pin 6.

Pin 3 and Pin 15 are used to regulate the brightness of the BPL (Back Plane Light). As mentioned above before the LCD operates on the principle of continuing or inhibiting the light passing through it; instead of producing light by itself. The light source comes from LED behind this LCD called BPL. Light brightness from BPL can be set by using a potentiometer or a trimpot. From Figure 6 Pin 3 (VEE) is used to regulate the brightness of BPL (by changing the current that enters BPL by using a potentiometers/a trimpot). While Pin 15 (BPL) is a Pin used for the sink of BPL LED.

Table 2 below is a summary of the function of each Pin of the 16 Pins of 2 × 16 LCD.

Table 2. Summary of functions of each Pin on a 2 × 16 LCD.

Pin	Name	Function
1	VSS	GND of the source voltage.
2	VCC	+5 V of the source voltage.
3	VEE	Adjust the brightness of the BPL by using an adjustable/variable resistor (potentiometer or trimpot).
4	RS	Register selector on the LCD, if RS = 0 then the selected register is an instruction register (the operation to be performed is a write operation/LCD configuration if Pin 5 (R/W¯) is given a logic 0), if RS = 1 then the selected register is a data register; if (R/W¯) = 0 then the operation performed is a data write operation to the LCD, otherwise if (R/W¯) = 1 then the operation performed is a read operation (data will be sent from the LCD to μC (microcontroller); it is usually used to read the busy bit/Busy Flag- BF of the LCD (bit 7/D7).
5	(R/W¯)	Sets the operating mode, logic 1 for reading operations and logic 0 for write operations, the information read from the LCD to μC is data, while information written to the LCD from μC can be data to be displayed or instructions used to configure the LCD. Usually, this Pin is connected to the GND of the power supply because we will never read data from the LCD but only write instructions to configure it or write data to the LCD register to be displayed.
6	Enable ¯	The LCD is not active when Enable Pin is either 1 or 0 logic. The LCD will be active if there is a change from logic 1 to logic 0; information can be read or written at the time the change occurs.
7–14	D0-D7	Data/instruction transmission channel between the LCD and the μC.
15	BPL	Sink for LED BPL (Back Plane Light). The VCC is connected to the LED anode and the cathode is connected to the BPL.
16	GND	Power supply ground. This Pin is connected to the power supply ground together with Pin 1 (VSS) of the LCD.

1.3 LCD Instructions

Just like the microcontroller, the LCD controller must also be programmed first to operate. Programming the LCD controller is an easy thing, all we have to do is only to give the appropriate logic level to the three Pins RS, R/W¯, E (Pins 4, 5, and 6), and DB0-DB7 Pin (Pin 7–14); see Figure 7. Each time we want to execute an instruction (read or write operation) there must be a change of logic level of Pin 6 (E-Enable); from logic 1 to logic 0 so that the instruction given to the controller can be executed properly, see Figure 7 below [12].

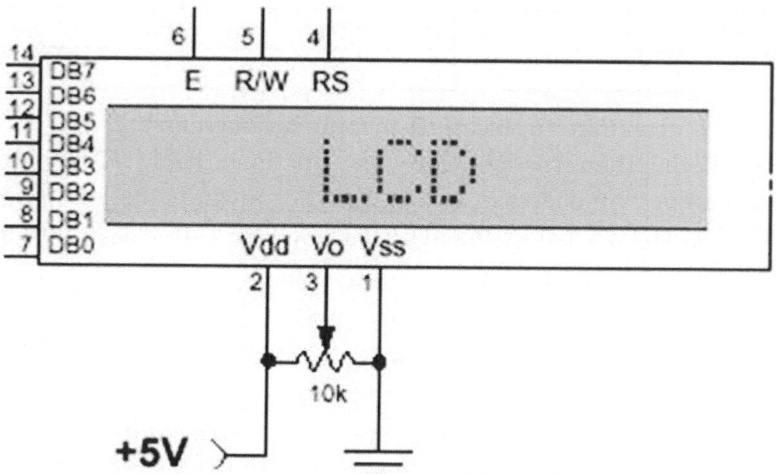

Figure 7. 2 × 16 LCD module [12].

The following are instructions for LCD controllers

Clear display

Clear display is an instruction that will clean the display and restore the cursor to the initial position (00h address). The format of this instruction is as follows,

RS	R/W‾	D7	D6	D5	D4	D3	D2	D1	D0
0	0	0	0	0	0	0	0	0	1

Note that RS = 0 which means we choose Instruction Register-IR and R/W‾ = 0 which indicates a write operation. Thus, the instruction we have to write on IR is 01h.

Home cursor

This instruction serves to return the cursor to the initial position (line 0, column 0) without changing the data on DDRAM. The format to do so is shown in the table below.

RS	R/W‾	D7	D6	D5	D4	D3	D2	D1	D0
0	0	0	0	0	0	0	0	1	X

D0(X) can be filled with either 0 or 1 (do not care) and in this case, we select X = 0. Thus, the instruction that must be written to return the cursor to the initial position on the IR is 02H [10, 13].

Entry mode set

This instruction will determine the direction of the cursor movement whether it is left or right and determines whether to shift the display or not. The format of this instruction is as follows.

RS	R/W⁻	D7	D6	D5	D4	D3	D2	D1	D0
0	0	0	0	0	0	0	1	I/D	S

Please pay attention to D1 (I/D-Increment/Decrement) it is a "bit" that determines whether the cursor position goes up or down. If D1 (I/D) = 0 (decrement), the cursor position will decrease, which will shift towards the left of one column. Conversely, if D1 (I/D) = 1 (increment), the cursor position will shift to the right of one column.

Likewise, the Bit D0(S) if D0(S) = 0 then the operation is normal, the cursor shifts but the display will stay in the same place, but if D0(S) = 1 then the display will shift one column to the right if D1 = 0 (the cursor will remain in the same place/ no movement. So, if we want a normal operating display and the cursor shifts to the right then the instructions, we have to give to IR is 06h and is 04h if you want to cursor shift to the left. Table 3 below summarizes the instructions given and the result of the cursor and the display movement.

Display ON/OFF

This instruction is used to ON or OFF the display, display or not display the cursor, and set the cursor to blink or not. The format of this instruction is as follows.

RS	R/W⁻	D7	D6	D5	D4	D3	D2	D1	D0
0	0	0	0	0	0	1	D	C	B

When D2 (D-Display) = 1 then the display will be on, but if D2 = 0 then the display will be off, just like in the case of D2, D1 (C-cursor) also has the same property if D1 = 1 then the cursor will be displayed otherwise (if D1 = 0) then the cursor will not be shown. D0 (B-blink) as mentioned above is used to set the cursor to blink or not if D0 = 0 then the cursor will not blink, but if D0 = 1 then the cursor will blink.

From Table 4 it can be seen clearly that the display will be ON when the instruction is 0Ch or bigger.

Function set

Function set (set data width, number of lines, and font size) with the format of instruction as follows,

RS	R/W⁻	D7	D6	D5	D4	D3	D2	D1	D0
0	0	0	0	1	DL	N	F	X	X

D4 (DL) ⇒ Data Length. If DL = 0, data width = 4 bits (D4-D7). If DL = 1, data width = 8 bits (D0-D7). D3 (N) ⇒ number of Line. If N = 0 number of lines = 1, if N = 1 number of lines is 2. D2 (F) = > font size. If F = 0 then the font size is 5 × 7 dot, if F = 1 then the font size is 5 × 10 dot. When data width of 4 bits is chosen, data must be sent twice.

Set CGRAM address

As mentioned earlier in Chapter 2 LCD controller, CGRAM (Character Generator RAM) is a part of CGROM (Character Generator ROM); See Figure 5 below [9]. From Figure 5 we can see that CGRAM with an address of 00000000b-00000111b (eight addresses) which is a part of CGROM is deliberately prepared by the manufacturer to be able to store custom use characters needed by the user and not be prepared by the manufacturer such characters as, battery picture, alarm picture, or other characters at the CGRAM addresses.

The format of instruction to program the CGRAM can be seen below.

RS	R/W̄	D7	D6	D5	D4	D3	D2	D1	D0
0	0	0	1	CGRAM address					

It is obviously seen that we need an instruction of 40h + the location of the CGRAM address where the pattern wants to be written to the Instruction Register-IR and followed by a write of the data to the Data Register (DR); this data pattern will be stored in the CGRAM address; the CGRAM is only 6 bits in length (D0-D5), thus the number of CGRAM addresses is only 64 addresses (address 0–63) [14]. To display a custom use character we need to save the patterns (8 patterns in maximum) into the 64-byte size memory location of CGRAM (0D-63D). The first pattern to be displayed will be referred to just like a normal CGROM address but the address is 00h. To save the pattern in the CGRAM address we need to run a command (40H + the first address of the pattern saved in the CGRAM locations) to the IR and followed by eight consecutives write data to the first eight locations of the 64D locations of the CGRAM. To display the CGRAM pattern we need to run the same instruction as to display the pattern in the CGROM, thus run the write command (80H + the DDRAM address where the pattern wants to be displayed on the LCD followed by the run of the write data instruction; the data is 00h. The same steps must be taken for the rest of the pattern saved in the CGRAM addresses (8D-63D); to study how to display a custom use pattern on the LCD in-depth one can visit the following site and download the paper, http://www.arpnjournals.org/jeas/research_papers/rp_2016/jeas_1016_5086.pdf [15].

Set DDRAM address

As mentioned before (above) to display one character on the specific location of the LCD we first have to write an instruction to IR; the instruction is "80h + the location of the character to be displayed on the LCD that has the associated addresses of DDRAM" which is on the controller before it can be displayed on the LCD; see the instruction format below.

After sending the mentioned above instruction to the Instruction Register-IR we need to write the data of the character (ASCII) to display on the LCD to the Data

Register; to display the character stored on the CGROM or CGRAM all we need to send is the location of the character stored on either CGROM or CGRAM instead of the code of the character. The format of the instructions to do that is as follows,

RS	$R/\overline{W}$	D7	D6	D5	D4	D3	D2	D1	D0
0	0	1	DDRAM address						

From the instruction format above we can see that the DDRAM address has a memory space of 128 locations (0000000b-1111111b) or 00H-7FH but the addresses that can be accessed only 80 locations from 00H-27H and 40H-67H each of them is 40 location memory spaces. For 2 × 16 LCD the addresses are limited to 32 locations with the first line DDRAM addresses of 00H-0FH (0D-15D) and the second line with the associated addresses range 40H-4FH (64D-79D).

3. Read busy flag and address

Before we can display data on the LCD, or read, change the data on the DDRAM first we must ensure the LCD module is in a not-busy state. The format to read busy Flag-BF of the LCD status is shown below,

RS	$R/\overline{W}$	D7	D6	D5	D4	D3	D2	D1	D0
0	1	0	0 0 0 0 0 0 1						

The results of the reading from busy flag will appear as shown in the table below.

D7	D6	D5	D4	D3	D2	D1	D0
BF (busy flag)	AC (address counter) contents						

From the table above we can see the busy flag-BF can be read on the 7th bit (D7), while the D0-D6 bit contains the address counter (cursor position). If D7 (BF) = 0 indicates the controller is not busy, on the contrary, if D7 (BF) = 1 indicates the controller is in a busy state. For example, the program to check busy bits or busy flag-BF can be seen in a published paper in the International Journal with the following website address, https://www.sciencedirect.com/science/article/pii/s1110016817300546 [16].

To write data to DDRAM or CGRAM

The format of the instruction to write data to DDRAM or CGRAM has the same instruction format as shown below. The operation is a writing data to the Data Register-DR; RS = 1 and R/W‾ = 0.

RS	R/W‾	D7	D6	D5	D4	D3	D2	D1	D0
1	0 DDRAM or CGRAM address								

Instruction format to read data from DDRAM or CGRAM has the same instruction format as shown in the table below,

RS	R/W‾	D7	D6	D5	D4	D3	D2	D1	D0
1	1	DDRAM or CGRAM address							

The operation is a read operation (R/W‾ = 1) of the content of DDRAM or CGRAM (RS = 1).

Table 3. Entry mode set, cursor movement, and display movement.

Instruction	Cursor movement direction	Display movement direction
04h	To the left one column	Stay in the same place/no movement
05h	Stay in the same place/no movement	Move to the right one column
06h	To the right one column	Stay in the same place/ no movement
07h	Stay in the same place/no movement	Move to the left one column

Table 4. Display, cursor, and blinking.

Instruction/command	Display	Cursor (display)	Cursor (blinking)
08h	OFF	NO	NOT
09h	OFF	NO	NOT
0Ah	OFF	NO	NOT
0Bh	OFF	NO	NOT
0Ch	ON	NO	NOT
0Dh	ON	NO	NOT
0Eh	ON	YES	NOT
0Fh	ON	YES	YES

1.4 LCD initialization

Before it can be used as a display for the MCS51 or other microprocessor or microcontroller, the LCD controller must first be configured or initialized by sending instructions to the Instruction Register-IR of the LCD controller. Before you can send instructions from the MCS51 to the LCD controller through P0 (P0.0-P0.7) we must choose the Instruction Register-IR on the LCD controller, by making RS = 0, resetting

R/W‾ (R/W‾ = 0), and make a change of signal level to the Enable ‾ Pin; from logic 1 to logic 0, see Table 2 for more details.

The LCD configuration process is to set the number of lines to be used (1 line or 2 lines), character font size (5 × 8 dot or 5 × 10 dot), the data width used (4 bits or 8 bits), and others. In this chapter, we only discuss the initialization of the LCD controller with a data width of 8 bits and 5 × 8 dot character font size.

We will develop some subroutines for this initialization process; to make it easier to use them in the future (just by calling the needed subroutines). The subroutines to be developed includes the write instruction subroutine, the write data subroutine, and the delay subroutine.

Because the initialization process is a write operation, we name this subroutine as a write_Insli subroutine. Before we build this subroutine, let us recall the Pins connection between the LCD module and MCS-51 for this initialization process.

For the initialization process the RS Pin must be reset (RS = 0) this is because we want to access the Instruction Register- IR to provide initialization instructions/ commands. RS Pin of the LCD controller is connected to P3.6 of MCS-51, for this reason, to reset this Pin (RS = 0) we can do it by running the CLR P3.6 statement.

As mentioned above initialization is a write operation, Pin R/W‾ must be reset (R/W‾ = 0) but due to the corresponding Pin being connected directly to the power supply GND we do not need to develop a subroutine to do that.

Enable‾ Pin of the LCD is connected to P3.7 of MCS-51; it must change from logic 1 to logic 0 to enable the process to be executed properly during this slot of time then we can write or develop some instructions for this process as follows,

SETB P3.7	; Enable ‾ (P3.7 = 1)
CALL delay	; call delay subroutine
CLR P3.7	; Enable ‾ (P3.7 = 0)

Thus, we can build the write_Insli subroutine as shown in subroutine write_Insli below.

write_instruction	:
MOV P0,r1	; fill P0 (D0-D7) with initialization instruction
	; which is in r1
CLR P3.6	;RS = 0 to select instruction register
SETB P3.7	; Enable ‾ get logic 1
CALL delay	; give appropriate delay in accord
	; with 2 × 16 LCD datasheet
CLR P3.7	; Enable ‾ get logic 0
RET	; return to the main program from the write_instruction subroutine

Based on datasheets from LCD 2 × 16 and the delay time needed to be around 1000 μs [17]. For that, we can build the subroutine delay as follows,

delay:		
	MOV r0,#0h	; fill r0 with 0
delay1:	MOV r7,#0fh	; fill r7 with 0Fh (15 s)
delay2:	DJNZ r7,$	; r7 = r7-1, if r7 is not equal to 0 jump to delay 2
	DJNZ r0,delay1	; r0 = r0-1, if r0 is not equal to 0 jump to delay 1
	RET	; return to the calling program or the main program

To analyze the delay time Table 5 below can be used [18, 19],

Table 5. Analyzing delay [18, 19].

	Instructions	**Number of MC-machine cycles**	**Time in seconds (1MC = 10^{-6} s)**
	MOV r0,#00h	1	10^{-6} s
delay1:	MOV r7,#0fh	1	10^{-6} s
delay2:	DJNZ r7, delay2	30 (15 × 2)	30 × 10^{-6} s
	DJNZ r0, delay1	512 (256 × 2)	512 × 10^{-6} s
	RET	2	2 × 10^{-6} s

From Table 5 above it can be seen clearly that the delay time is 546 MC (546 × 10−6 s) or can be written as follows 546 μs; you can adjust the content of register r0 and r7 to obtain the recommended 1000 μs (enable cycle time) to ensure the LCD is really ready to receive the next command (not in a busy state).

To write data to the LCD is a very similar process to write instruction to the LCD (Instruction Register-IR) as shown in the write_inst subroutine; the only difference between them is the logic state of RS; instead of RS = 0 we give RS a logic 1 (RS = 1) for this process; write data to the Data Register-DR. The subroutine to write data to the Data Register-DR is called write_data as shown below,

write_data:	
MOV P0,r1	; fill in P0 (D0-D7) with the data you want to display
	; which is in r1
SETB P3.6	;RS = 1 to select data register
SETB P3.7	; Enable ˉ get logic 1
CALL delay	; call delay subroutine to give a delay in accord
	; with 2 × 16. LCD datasheet
CLR P3.7	; Enable ˉ get logic 0
RET	; return from subroutine write_instruction

From the 2 × 16 LCD control data [20], there are some things we have to do for the initialization process, namely

❖ Function set (set the width of the data, number of lines, and character font size) with the following instructions format

D7	D6	D5	D4	D3	D2	D1	D0
0	0	1	DL	N	F	X	X

NB:

DL ⇒ Data Length. If DL = 0, data width = 4 bits (D4-D7).

If DL = 1, data width = 8 bits (D0-D7).

N ⇒ number of line. If N = 0 the number of lines is 1, if N = 1 the number of lines is 2.

F ⇒ font size. If F = 0 then the font size is 5 × 8 dot, if F = 1 then the font size is 5 × 10 dot.

If we choose the data width of 4 bits, the process of sending or receiving data or instruction must be done twice. On this occasion we only discuss the 8-bit data width (Data Length-DL) with (DL = 1), the font size 5 × 8 dot (F = 0), the number of lines 2 (N = 1), bits D1, and D0 are filled with logic 0; therefore the instruction that must be given to set the function set is 00111000B = 38H and we can use the instructions or statements below to do the function set.

MOV r1,#38h;	fill r1 with instruction (function set) 38h	
CALL write_instruction;	CALL write_instruction	

❖ Entry mode set, this initialization instruction has the following instructions format

D7	D6	D5	D4	D3	D2	D1	D0
0	0	0	0	0	1	I/D	S

Note Below (NB):

D1 (I/D) ⇒ Increment/decrement, if D1 (I/D) = 0 this bit will tell the controller to move the cursor to the left one column (D1 = 0) or move to right one column (D1 = 1) on condition D0(S) =0. But if D0(S) =1 the whole display (the cursor and the character or characters) will move to the left one column when D1 = 1 and to the right one column when D1 = 0; this instruction also increases or decrease the DDRAM address based on the D1 and D0 status. If D0 = 1 and D1 = 0 the DDRAM address will increase one column, but in the contrary if D0 = 1 and D1 = 1 the DDRAM address will decrease one column.

In this book, we will set S and I/D thus the instruction we send to configure the Entry Mode Set is 0000 0111b (07h). The instructions for initialization Entry Mode Set of the LCD we can use the statements below,

MOV r1,#07h;	fill r1 with instructions (entry mode set) 07h
CALL write_instruction;	call the write_instruction subroutine

Display on/off cursor, these instructions have a format as shown below

D7	D6	D5	D4	D3	D2	D1	D0
0	0	0	0	1	D	C	B

This instruction is used to on or off the display, display or does not display the cursor, and set the cursor to blink or not. When D2 (D-Display) = 1 then the display will be on the opposite if D2 = 0 then the display will be off, so is with the case of D1 (C-cursor) if D1 = 1 then the cursor will be displayed otherwise if D1 = 0 then the cursor will not be shown. D0 (B-blink) as mentioned above is used to set the cursor to blink or not to blink if D0 = 0 then the cursor will not blink, but if D0 = 1 then the cursor will blink. To initialize the on /off cursor display we use the initialization of display on, cursor off, and blink off thus the instructions that must be sent to IR for it is (0Ch). The instructions used for the initialization of the display on/off cursor are shown below.

MOV r1,#0Ch;	fill r1 with instruction (display on/off cursor) 0Ch
CALL write_instruction;	call the write_instruction subroutine

Clear display initialization instructions to clean the display; it has an instruction format as shown in the table below.

D7	D6	D5	D4	D3	D2	D1	D0
0	0	0	0	0	0	0	1

Thus for the initialization process to clean the display, the instruction that must be given to IR is (01h) and can be done by running the instructions below.

MOV r1,#01h	; fill r1 with instruction (clear display) 01h
CALL write_instruction	; call the write_instruction subroutine

Shift cursor and display, this initialization instruction is used to shift the cursor position or the entire display to the right and to the left. The format of instructions for this operation is as follows

D7	D6	D5	D4	D3	D2	D1	D0
0	0	0	1	S/C	R/L	X	X

The conducted operation is based on the logic status (logic level) of both bits D3 (S/C) and D2 (R/L) as shown in the table below (Table 6) [6, 14],

Table 6. Shift cursor and dispaly [6, 14].

D3 (S/C)	D2 (R/L)	Explanation
0	0	Shift the cursor position to the left
0	1	Shift the cursor position to the right
1	0	Shift the whole display to the right
1	1	Shift the whole display to the left

1.5 More instructions

The following are other commands besides the initialization command, write data, and write instructions mentioned above. These instructions or commands determine the Display Data RAM (DDRAM) address, read the status of the LCD (Busy Flag-BF), and specify the Character Generator RAM (CGRAM address).

1.5.1 Specifying DDRAM addresses

As mentioned earlier that each location on the LCD has an address that corresponds to the address on DDRAM; See Chapter 2 LCD Controller. To display a character on the LCD you must write an instruction (80h + the address location on the LCD where the character wants to be displayed) to the Instruction Register followed by writing the data (the character code you want to display) to the Data Register-DR.

Each time we want to read or change data on certain addresses on DDRAM we must write an instruction to the Instruction Register-IR to be able to access the address before carrying out the activities mentioned above (displaying characters on the LCD, read data on certain addresses from the DDRAM, and change data at the address). The format of the instruction for it is shown as follows,

D7	D6	D5	D4	D3	D2	D1	D0
1	Address ofDDRAM						

From the instruction format above we can see that the instructions we have to write to the Instruction Register-IR is 80H + the DDRAM addresses where the character is to be displayed; we also have to send signals (RS = 0, R/W¯ = 0). For example, to put the cursor on the first line of the second column (on the LCD then we must provide the following instructions,

MOV r1,#82h	; fill r1 with 82h (80h + 02h)
CALL write_instruction	; call the write_instruction subroutine

Likewise, if we want to put the cursor in the position of the first line of column 15 (0fh) then we must provide the following instructions, [10, 13].

MOV r1,#8Fh	; fill r1 with 8Fh (80h + 0Fh)	
CALL write_instruction	; call the subroutine write_instruction	

1.5.2 Reading LCD busy status

Before we display data on the LCD, or read, change data on the DDRAM first we must ensure the LCD module is in a non-busy state. Format to read busy LCD status is shown below,

D7	D6	D5	D4	D3	D2	D1	D0
BF (busy flag)	AC (address counter) contents						

To be able to read the busy status of this LCD module we must reset RS (RS = 0), and set R/W¯ (R/W¯ = 1). As shown in the format of the busy/not busy of the status of LCD Bit 7 is the BF (busy flag) and bits 0- bit 6 are the contents of the Address Counter. In this book we will never read BF status; It is assumed that the LCD module remains on standby [6, 12, 15]. Most of the time people use a delay loop to make sure that LCD is ready not in a busy state; this is due to the grounded of Pin 5 (R/W¯) of the LCD to the ground (GND) of the power supply. If you want to study how to read the BF status of the LCD module you can download and study a research paper published on the following site, https://www.sciencedirect.com/science/article/pii/S1110016817300546 [16].

1.5.3 Specifying the CGRAM address

To be able to read or write data on the CGRAM (Character Generator RAM), we must first determine the CGRAM address to be accessed. The format to access CGRAM can be seen below (RS = 0, R/W¯ = 0); a write to the Instruction Register of the LCD with the command/instruction 40h.

D7	D6	D5	D4	D3	D2	D1	D0
0	1 CGRAM address						

The instructions below can be used to determine the CGRAM address (10H = 00010000B) [6, 12].

MOV r1,#01010000b	; fill r1 with (40h + 10h = 50h) to determine the CGRAM; address
	; at address 10h
CALL write_instruction	; call the write_instruction subroutine

From the format to access the addresses of the CGRAM we can see that the address of the CGRAM which is possible to access is from 000000b to 111111b or if written in the format of the hexadecimal is 00h to 3fh and in the format of decimal is 0d to 63d, thus the total address of the CGRAM is 64 bytes.

To save one custom pattern character with a size of 5 × 8 dot (5 columns × 8 lines) it takes eight address locations on CGRAM, thus 64 locations on CGRAM can only accommodate or store eight custom patterns; 1 memory location is occupied by 1 custom pattern line. If we make a storage memory map of CGRAM we can see it as shown on the CGRAM memory map below.

Pattern no.	**CGRAM address**	
	Hexa decimal	**Decimal**
1	00-07	0–7
2	08-0F	8–15
3	10-17	16–23
4	18-1F	24–31
5	20-27	32–39
6	28-2F	40–47
7	30-37	48–55
8	38-3F	56–63

For example, to create the character of the letter "T" then this data can be stored in eight consecutive CGRAM addresses (Table 7) [6].

Table 7. Graph of the formation of a custom pattern for the letter T with a size of 5 × 8 dots [21].

	Custom pattern							
	1	**8**	**4**	**2**	**1**		**Hexa**	**Decimal**
Row 1							1F	31
Row 2							04	4
Row 3							04	4
Row 4							04	4
Row 5							04	4
Row 6							04	4
Row 7							04	4
Row 8							04	4

And to make characters with alarm images can be seen in Table 8 below,

If the custom pattern character size used is 5 × 10 dot then CGRAM will only be able to accommodate 6 custom pattern characters. Examples of programs to display special characters on the LCD can be seen in a published paper on the International Journal with the following website address, http://www.arpnjournals.org/jeas/research_papers/rp_ [15].

Table 8. Graph of the formation of a custom pattern alarm image with a size of 5 8 dots.

	Custom pattern						
	1	8	4	2	1	Hexa	Decimal
Line 1						04	4
Line 2						0E	14
Line 3						0E	14
Line 4						0E	14
Line 5						0E	14
Line 6						1F	31
Line 7						04	4
Line 8						00	0

1.5.4 To return the cursor to the initial position

To restore the cursor to the initial position (line 0, column 0) without changing the data on DDRAM. The format to do this is as follows (RS = 0, R/W̄ = 0), D0 is marked with X we can fill it with either 0 or 1 (do not care) and we choose X = 0. Thus the instruction that must be written to the IR (Instruction Register) is 02H [10, 13].

D7	D6	D5	D4	D3	D2	D1	D0
0	0	0	0	0	0	1	X

The instructions below will return the cursor to the starting position (home).

MOV r1,#02h	; fill r1 with immediate data 02h
CALL write_instruction	; call the write_instruction subroutine

1.6 LCD initialization subroutine

As mentioned earlier, before the LCD can be used it must be initialized/ configuration initially. After studying the instructions for the initialization process above, we will build a subroutine for the initialization process and call it Init_LCD. In developing this subroutine all that needs to be done is to combine the entire initialization instructions that have been discussed previously. The init_LCD subroutine can be seen as follows,

init_LCD:	
MOV r1,#01h	; clear display
CALL write_instruction	
MOV r1,#38h	; function set, data size of 8 bit, 2 lines,
	; font size 5 × 8 dot

CALL write_instruction	
MOV r1,#0Ch	; display ON, cursor OFF, blink OFF
CALL write_instruction	
MOV r1,#06h	; entry mode set, the display will not ; move, but the cursor will ; move to the right one column
CALL write_instruction	
RET	

1.7 Displaying a character on the LCD

The instruction or command to display a character on the LCD is 80h. To display a character at a certain location we must add 80H with the location address on the LCD DDRAM where the character is to be displayed. For example, if we want to display a character in the first row and the second column on the LCD we must add 80h + 01h = 81h, while to display the character on the second line of column 4 then we must write 80H + 43H = C3H to the Instruction Register-IR.

The program below is an example of a program to display character A on the second line of column 4 (C3H) [21].

	Org 0h	
	CALL init_LCD	; initialization of the LCD
Start:		
	MOV r1,#C3h	; fill r1 with C3h, put the character
		; on row 2 column 4 of the LCD
	CALL write_instruction	; call write_instruction
	MOV r1,#'A'	; fill r1 with 'A' the character
		; to be displayed on the LCD
	CALL write_data	; call write_data
	SJMP $	; halt or stop

The paper that can be used as a learning media for reading a 4 × 4 keypad and displaying the results of typing characters on the LCD can be seen at the following site address, http://www.arpnjournals.org/jeas/research_papers/rp_2018/jeas_0418_7024.pdf [22].

1.8 Displaying more than 1 character on the LCD

To display more than one character is almost the same as to display a single character. The program below is an example of a program to display seven characters "MANDATE" in the first line of the first column (80h) [21, 23].

	Org 0h	
	CALL init_LCD	; LCD initialization
start:		
	MOV r1,#80h	; fill r1 with 80h, put the character
		; on LCD row 0, column 0
	CALL write_instruction	; call write_instruction
	MOV r1,#'M'	; fill r1 with 'M' the character
		; to be displayed on the LCD
	CALL write_data	; call write_data
	MOV r1,#'A'	; fill r1 with 'A' the character
		; to be displayed on the LCD
	CALL write_data	; call write_data
	MOV r1,#'N'	; fill r1 with 'N' the character
		; to be displayed on the LCD
	CALL write_data	; call write_data
	MOV r1,#'D'	; fill r1 with 'D' the character
		; to be displayed on the LCD
	CALL write_data	; call write_data
	MOV r1,#'A'	; fill r1 with 'A' the character
		; to be displayed on the LCD
	CALL write_data	; call write_data
	MOV r1,#'T'	; fill r1 with 'T' the character
		; to be displayed on the LCD
	CALL write_data	; call write_data
	MOV r1,#'E'	; fill r1 with 'E' the character
		; to be displayed on the LCD
	CALL write_data	; call write_data
	SJMP $	; halt or stop

1.9 A 4-bit mode 2 × 16 LCD module

Figure 8 below shows a 4 bit mode 2 × 16 LCD module. Unlike the LCD in 8 bit mode in 4-bit mode only used 6 Pins as the LCD interface to other equipment (a microcontroller, a microprocessor, or a computer), 4 bidirectional data bus (D4-D7) (Pin 11-Pin 14), and 2 control signal RS (Pin 4) and E (Pin 6) [6, 24].

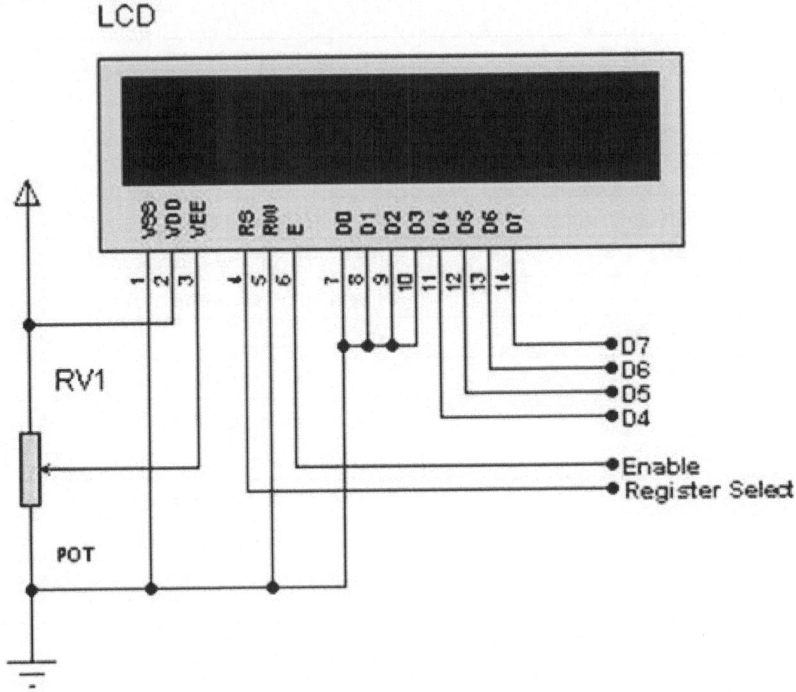

Figure 8. 4 bit mode 2 × 16 LCD module [6, 24].

Data channels (Pin 7–Pin 10) D0-D3 are not used and connected to the ground of the power supply. Note the Pin (R/W¯) is connected to the ground, thus in this mode, one can not read the status or data from the module all you can do only to write either instruction or data to the appropriate register of the LCD controller. If you want the status of the LCD and the data in it to be read then Pin (R/W¯) must be connected to another of the controller that can provide a high signal for a read operation or a low signal for writing operations.

The advantage of using LCD in 4-bit mode compared to 8-bit mode is the number of Pins used as interfaces with microprocessors, microcontrollers, or computers is lesser, but also it has some shortcomings in programming; more complicated (data transmission must be done twice, the first is the high nibble data (D4-D7) then followed by sending low nibble (D0-D3) through D4-D7 data bus. Because the channels used are only 4 bits (D4-D7), a shift operation must be carried out to shift data from low nibble to high nibble (D4-D7) or swap instruction that exchanges between high nibble and low nibble.

1.10 Summary

LCD is an alternative display to the Light Emitting Diode-LED and seven segment display which are commonly used in various microcomputer and microcontroller applications or projects. LCDs are available in various sizes; size is usually based on the number of rows and columns. LCD is basically a module consisting of a display and controller. The controller is a microcontroller that has two types of memory (the DDRAM and the CGROM with CGRAM resides in it; CGROM is a non-volatile memory but not the CGRAM, it is a volatile memory) and two registers (the instruction register and the data register) in it. The LCD modules that are commonly used are LCD with a size of 2 × 16; has two lines and each line can display 16 characters.

The memory in the controller can be divided into two, the Display Data RAM-DDRAM, and Character Generator ROM-CGROM. DDRAM is a RAM memory (volatile memory) each address on the DDRAM has a corresponding address to each location on the LCD; where the character will be displayed. LCD with a size of 2 × 16, the first line has the DDRAM addresses corresponding to the LCD addresses of 00h-0Fh, while the second line has a DDRAM address corresponding to the LCD addresses of 40h-4Fh.

CGROM is a non-volatile memory where various characters are permanently stored in it by the manufacturer, see Figure 5 to be displayed on the LCD by sending the address where the character is stored in the CGROM of the LCD controller; The command (80h + column in the first line where the character is to be displayed) is written to the LCD controller's instruction register.

To display the character on the second line, command C0h + column on the second line where the character is to be displayed in the LCD controller instruction register. The LCD controller manufacturer has also prepared a special location 00h-07h on the CGROM is a place to store special characters that are not provided by the LCD controller manufacturer. This special character is built by the user by sending a 40h command to the instruction register and sending the next eight bytes (the graph of the constructed character) to the data register eight times in sequence.

LCD can be operated in two modes, namely eight-bit mode and four-bit mode. In the eight-bit mode, the eight data busses of the LCD controller are connected to the data bus of the microcontroller used to communicate between the two. In 4-bit mode, only four channels of the 8-bit width data bus of both LCD controllers and microcontrollers are used, namely D7-D4; taken from the data bus of the LCD controller, D3-D0 is not used and is connected to the power supply ground (GND). The eight-bit mode does use more wires than the four-bit mode, but is simpler in programming; the four-bit mode requires fewer wires but is more complicated to program and requires sending instructions and data twice.

In order to use the LCD module, it must be programmed first. By programming it, we can display the characters we want to display on the LCD. To enable it to display the characters, the LCD must first be configured/initialized first. Initialization is a process where we tell the LCD controller to display the characters on the LCD in the mode we want; font type, shift the cursor or the display to the left or to the right direction, blinking cursor or not, characters are displayed on the desired LCD row and column and so on. Basically, LCD programming consists of two operations, namely read and write operations; write or read the instructions to or from the instruction registers-IR or write the data to the data registers-DR. The write operations are used more than the read operation. The read operation is usually used to find out whether the LCD is busy or not (by reading the LCD controller's busy flag-BF); if the LCD is in a busy state, sending or reading the instructions/data to or from it will be useless. If you have connected Pin (R/W¯) to the power supply ground and you could not read the busy flag-BF status anymore, you can use the delay loop time suggested by the LCD datasheet instead.

Glossary

CGROM CGROM stands for Character Generator ROM; it is a ROM of the LCD module. This ROM is filled with 208 Dot Character Pattern 5 _ 8 and 32 dot character pattern 5 _ 10 that have been filled during its fabrication; Cannot be modified anymore. The address of the characters stored on this ROM will be sent to a certain memory location on DDRAM to be displayed on the LCD. CGRAM CGRAM stands for Character Generator RAM; is one type of RAM of the LCD module. This RAM is basically a part of CGROM which is intentionally emptied by its manufacturer in order to be used by users to build special characters that are not prepared by the manufacturer to be displayed on the LCD (DDRAM); contains a point character made by the user; custom use character. There are eight characters with a 5 _ 8 format and six characters with a 5 _ 10 format. This memory is a volatile memory where the data will be lost if the energy source is removed from it; LCD does not get a voltage source. DDRAM DDRAM stands for Display Data RAM. The DDRAM is a RAM used to save for a while the 8-bit character that will be displayed on the LCD; Both characters from CGROM and characters built by the user themselves and stored in CGRAM; The stored characters in CGRAM are volatile (will disappear if the power source is removed from it). The LCD with a size of 2 lines _ 16 Characters, the address of the DDRAM for the first line is 00H-0FH and for the second line is 40H-4FH.

BPL BPL is an abbreviation of the Back Plane Light, it is the Sink or Ground of the light source for the LCD, in this case, is a LED; Cathode (_) Lead of the LED is connected to the BPL Pin and LED Anode (+) Lead is connected to the VCC voltage source. Light emitted by the LED is used to illuminate the displayed character on the LCD. Character brightness displayed on the LCD can be set up by using a variable resistor such as a potentiometer or a trimpot.

Write operation it is an operation to place data/instructions from outside the LCD controller to each register (the data register and the instruction register) inside the LCD controller to be further processed by the LCD controller; data to be displayed on the LCD and instructions to be executed.

Read operation it is an operation to read the busy bit of the LCD; to read the content of the instruction register. The read operation is generally used to read the status (busy/not busy) of the LCD. If the LCD is busy we send neither data nor instructions to it; data/instructions will be lost if we send them to LCD. Initialization it is a program that tells the LCD controller to set up the characteristics of the characters that we want to display on the LCD; set the number of lines of the LCD that we will use (1 line or 2 lines), the character font size (5 _ 8 or 5 _ 10); 5 _ 8 (5 columns _ 8 rows), 5 _ 10 (5 columns _ 10 rows), the LCD controller data width used (4 bits or 8 bits) and others. The busy bit the busy bit is the 7th bit of the LCD controller data bus. We need to read the 7th bit (Busy Flag-BF) before giving further data or instructions to the LCD controller. This 7th bit will be a logic 1 (High) if it is busy and a logic 0 (Low) if it is not busy. The operation to read the 7th bit (Busy Bit) is an operation to read the contents of the Instruction Register IR; bit 7 in this case [22].

References

1. Difference between Active Matrix LCD and Passive Matrix LCD. Available from: https://www.geeksforgeeks.org/difference-between-active-matrix-lcd-and-passive-matrix-lcd
2. Passive LCD. Available from: https://www.usmicroproducts.com/displays/custom-passive-lcd
3. LCD block diagram. Available from: https://lh3.googleusercontent.com/0mMcFMaYqEzSjY2YLvcSSMR-NRy-xYPsUcKEMJoj_8iZCxG20T9cBA2ZTayD7d5ISJcrgA=s85
4. The types and the size of the LCD. Available from: https://www.engineersgarage.com/character-lcd-pinout-working
5. Interfacing LCD to 8051. Available from: https://www.dnatechindia.com/images/stories/8051_Tutorial/Interfacing_LCD_to_8051_/2x16_Line_Alphanumeric_LCD_Display.jpg
6. Poltak Sihombing DS. A-Z Microcontroller 8051 Perangkat keras, Antar Muka, Pemograman dan Aplikasi. Medan: USU Press; 2019
7. LCD Addressing. Available from: http://web.alfredstate.edu/faculty/weimandn/lcd/lcd_addressing/lcd_addressing_index.html
8. Two line LCD. Available from: https://lh3.googleusercontent.com/IgFjlJwrahfBNE4l4-6_C44w-YB_rLEMhr8ox4qvdhQaTGLolft7ZHNEZY9hSV4FeaUm = s170
9. LCD16 × 2 CGROM and CGRAM. Available from: https://www.handsonembedded.com/wp-content/uploads/2018/04/char_codes.png
10. LCD Commands HD44780 instruction set. Available from: https://mil.ufl.edu/3744/docs/lcdmanual/commands.html
11. LCD character 2 × 16. Available from: https://www.mytutorialcafe.com/image/gambar41.gif

12. The Schematics of the LCD. Available from: https://www.microcontrollerboard.com/lcd.html
13. Commands and Instruction set. Available from: https://www.8051projects.net/lcd-interfacing/commands.php
14. LCD Commands. HD44780 instruction set. Available from: https://mil.ufl.edu/3744/docs/lcdmanual/commands.html#Sca
15. Mhd Furqan DRPS. Developing a teaching media of microcontroller 8051 in displaying CGRAM character on LCD by using the MCU 8051 IDE and ASM51 in supporting ALFHE. ARPN Journal of Engineering and Applied Sciences. 2016;11(19):11363-11367. Available from: http://www.arpnjournals.org/jeas/research_papers/rp_2016/jeas_1016_5086.pdf
16. Dahlan Sitompul PS. A teaching media of using busy bit and SDCC in displaying character string on LCD in MCU 8051 IDE. Alexandria Engineering Journal. 2018;57(2):813-818. Available from: https://www.sciencedirect.com/science/article/pii/S1110016817300546
17. Bus timing characteristics. p. 49. Available from: http://lcd-linux.sourceforge.net/pdfdocs/hd44780.pdf
18. 8051/8052 Microcontroller Tutorial. 8051/8052 Instruction Set. Available from: https://www.hobbyprojects.com/8051_tutorial/8052_instruction_set.html#is
19. Philips 80C51 Family Programmer's Guide. Available from: https://www.keil.com/dd/docs/datashts/philips/p51_pg.pdf
20. All about circuits. How to Interface a 16×2 LCD Module with an MCU. Available from: https://www.allaboutcircuits.com/technical-articles/how-to-a-162-lcd-module-with-an-mcu/
21. 8051 Microcontroller Assembly Language Programming. Available from: https://www.electronicshub.org/8051-microcontroller-assembly-language-programming/
22. Dahlan Sitompul SDF. An active learning media of interfacing microcontroller 8051 to 4 × 4 keypad using mcu 8051 IDE and ASM-51. ARPN Journal of Engineering and Applied Sciences. 2018;13(8):2987-2992. Available from: http://www.arpnjournals.org/jeas/research_papers/rp_2018/jeas_0418_7024.pdf
23. Microprocessors & Microcontrollers 8051 Assembly language Programming. Available from: https://amiestudycircle.com/free-samples/amie/chapters/8051-programming-sec-b-elect-m&m.pdf
24. LCD connections in 4-bit mode. Available from: https://www.8051projects.net/lcd-interfacing/lcd-4-bit.php
25. The 8051 Simulator for Teachers and Students. Available from: https://www.edsim51.com/
26. MCU 8051 IDE. Available from: https://sourceforge.net/projects/mcu8051ide/

Index

V

W

9 789360 845834

Printed at Repro India Ltd.